PLANNING FOR EMPIRE

Studies of the Weatherhead East Asian Institute, Columbia University

The Weatherhead East Asian Institute is Columbia University's center for research, publication, and teaching on modern and contemporary East Asia regions. The Studies of the Weatherhead East Asian Institute were inaugurated in 1962 to bring to a wider public the results of significant new research on modern and contemporary East Asia.

Planning for Empire

Reform Bureaucrats and the Japanese Wartime State

Janis Mimura

CORNELL UNIVERSITY PRESS, ITHACA AND LONDON

First published 2011 by Cornell University Press
First printing, Cornell Paperbacks, 2017

Library of Congress Cataloging-in-Publication Data

Mimura, Janis, 1963–
    Planning for empire : reform bureaucrats and the Japanese wartime state / Janis Mimura.
        p. cm. — (Studies of the Weatherhead East Asian Institute, Columbia University)
    Includes bibliographical references and index.
    ISBN 978-0-8014-4926-0 (cloth : alk. paper)
    ISBN 978-1-5017-1354-5 (pbk. : alk. paper)
    1. Japan—Politics and government—1926–1945.   2. Bureaucracy—Japan—
History.   3. Technological innovations—Japan—History.   4. Fascism—
Japan—History.   5. Civil-military relations—Japan—History.   6. Manchuria
(China)—History—1931–1945.   I. Title.   II. Series: Studies of the Weatherhead
East Asian Institute, Columbia University.
    DS888.5.M546   2011
    952.03'3—dc22   2010037180

For my father, Takuo Mimura,
and in memory of my mother, Akiko Mimura

Contents

Acknowledgments

I am grateful for the assistance that I have received from numerous individuals and institutions in completing this study. First and foremost, I thank my teachers at Berkeley, Andrew Barshay and Irwin Scheiner, whose guidance, enthusiasm, and support sustained me in the long road to this book. In Japan, I am indebted to Kobayashi Hideo, Hara Akira, Imura Tetsuo, Itō Takashi, Kawahara Hiroshi, and Kudō Akira for generously sharing their time, advice, and insights on wartime Japan. Masayo Takashi helped me access valuable Manchuria archives, and Kobayashi Hideo, Imura Tetsuo, and Satoshi Sasaki shared some of their own primary documents on wartime planning. In the process of research and writing I benefitted from the advice and assistance of Katalin Ferber, Louise Young, Gerald Feldman, Michael Barnhart, Miles Fletcher, Bai Gao, Mark Metzler, Tatsushi Hirono, Sharon Holt, and Donna Rilling. The book was considerably improved by the detailed comments and suggestions of Larry Frohman, Richard Smethurst, and two anonymous readers from Cornell University Press. I thank Roger Haydon at Cornell University Press for his expert guidance and Jack Rummel for copy editing the manuscript. I also wish to thank Carol Gluck, Kim Brandt, and the staff at the Weatherhead East Asian Institute for their assistance in publishing this book.

Research and writing were made possible by grants from the Fulbright Foundation; University of California at Berkeley, Department of Education Foreign Area Language Studies Program; the New York State United University Profession Drescher Program; and the Deutscher Akademischer Austauschdienst. Kirsten Refsing and the Department of Japanese Studies at the University of Hong Kong provided me with a congenial place to write. Carl Steenstrup and Franz Waldenberger arranged for my affiliation as a visiting scholar at the University of Munich. I warmly thank Kobayashi Hiroko, Miyahara Hiromichi, Susan Barclay, John Lee, Pam Burdman, Allison Rottman, George Mimura, and Laura Mimura for their assistance during my various travels between the United States, Germany, and Japan. Finally, I cannot adequately express my gratitude for the patient and loving support of my husband, Martin Kock, and daughters, Karina and Vivien.

INTRODUCTION

In September 1931, officers of Japan's Kwantung Army blew up train lines of the South Manchuria Railway Company, or Mantetsu, and seized Manchuria. What began as an unauthorized military scheme to secure resources for a future total war evolved into an ambitious imperialist project to create the Manchurian state of Manchukuo, a domestic New Order, and a Greater East Asia Co-Prosperity Sphere. This book begins with the Manchurian occupation not to analyze Japanese militarism and the history of Japan's fifteen-year war of military aggression that ended in 1945. Rather, it seeks to examine the history of Japan's transwar embrace of technocratic rule and the ambitious drive for power of a group of civilian planners known as the "reform bureaucrats" (*kakushin kanryō*). During the war, these bureaucrats laid the foundations for a new type of state and society dominated by technocrats. After the war, they drew on the lessons and experiences of wartime planning and their prewar institutional and personal planning networks to realize their political ambitions and vision.

The leaders of Manchukuo were neither fanatic militarists nor manipulated leaders. They were highly rational and conscientious public servants who promoted a vision of an ultramodern Japan. They represented a new breed of managers who were dedicated to the concept of the organization and its ethic of collective expertise and cooperative action. On the one hand, they embraced Western science and technology, but on the other, they rejected the liberal capitalist system of free enterprise based on the principles of private property, private profit, and business autonomy. Most were sent to Manchuria as junior or midlevel specialists and honed their planning skills in developing Manchuria's economy. This Manchurian ruling clique was popularly known as *ni-ki, san-suke,* or "two ki's and three suke's"—a reference to the last syllables in their first names. The two "ki"s were Tōjō Hideki, Kwantung Army chief of staff and military police chief, and Hoshino Naoki, director of general affairs in Manchukuo. The three "sukes" were Kishi Nobusuke, vice director of general affairs and vice minister of industry in Manchukuo; Ayukawa Yoshisuke, head of Nissan and the Manchurian Heavy Industries Corporation, or Mangyō; and Matsuoka Yōsuke, Mantetsu president. On returning to Japan in the late 1930s, most occupied the top ministerial posts within the Japanese wartime government. Tōjō served as Japan's prime minister from 1941 until 1944, during which time he also held the army, home, and munitions portfolios. Kishi was appointed commerce minister and vice minister of munitions, Hoshino served as director of the cabinet planning board and chief cabinet secretary, and Matsuoka became foreign minister.[1] From their positions of authority they became the official proponents of fascism. They were but the most prominent representatives of a wider group of professionals who embraced a fascist vision of Japanese hegemony in Asia.

Of the five leaders, Kishi Nobusuke had the most profound impact on Japanese wartime planning. Kishi embraced a bold vision of Japan as a technocratic superpower.[2] He believed that only leaders with managerial ability and expertise should govern and had little patience for privileged, old-school elites. At the same time, Kishi was a pragmatist and a skilled and tireless mediator who brought together the military, business, right-wing activists, and left-wing planners. He was not a popular, charismatic leader but a behind-the-scenes deal-maker who could steer groups toward a middle ground and blunt the sharp edges of conflicting agendas. Kishi was loyal to his followers, delegated important tasks to talented

---

1. English-language studies of these leaders include Robert C. Butow, *Tōjō and the Coming of the War* (Stanford: Stanford University Press, 1961); Haruo Iguchi, *Unfinished Business: Ayukawa Yoshisuke and U.S.–Japan Relations, 1937–1953* (Cambridge, MA: Harvard University Asia Center, 2003); David Lu, *Agony of Choice: Matsuoka Yōsuke and the Rise and Fall of the Japanese Empire, 1880–1946* (Lanham, MD: Lexington Books, 2002); Richard J. Samuels, *Machiavelli's Children: Leaders and Their Legacies in Italy and Japan* (Ithaca: Cornell University Press, 2003); Chalmers Johnson, *MITI and the Japanese Miracle* (Stanford: Stanford University Press, 1982).

2. For recent scholarly works on Kishi in Japanese, see Hara Yoshihisa, *Kishi Nobusuke: Kensei no seijika* (Tokyo: Iwanami shoten, 1995); Hara Yoshihisa, *Kishi Nobusuke shōgenroku* (Tokyo: Mainichi shinbunsha, 2003); Nakamura Takafusa and Miyazaki Masayasu, *Kishi Nobusuke seiken to kōdo seichō* (Tokyo: Tōyō keizai shinpōsha, 2003). Other works on Kishi include Ushio Shiota, *Kishi Nobusuke* (Tokyo: Kōdansha, 1996); Hosokawa Ryūichirō, *Kishi Nobusuke* (Tokyo: Jiji tsūshinsha, 1986); Tajiri Ikuzō, *Shōwa no yōkai: Kishi Nobusuke* (Tokyo: Gakuyō shobō, 1979). Postwar writings and interviews by Kishi include Kishi, Yatsugi, and Itō, "Kankai seikai rokujūnen: Dai-ikkai Manshū jidai," *Chūō kōron* (September 1979); Kishi, Yatsugi, and Itō, "Shōkō daijin kara haisen e," *Chūō kōron* (October 1979); Kishi, *Waga seishun* (Tokyo: Kōsaidō, 1983); Kishi, *Kishi Nobusuke kaikōroku* (Tokyo: Kōsaidō, 1983); Kishi Nobusuke, Yatsugi Kazuo, and Itō Takashi, eds., *Kishi Nobusuke no Kaisō* (Tokyo: Bungei shunjū, 1981).

junior staff, and took big risks. These qualities served him well and enabled him to flourish in the turbulent years of early postwar Japan. After serving three years in Sugamo Prison as an unindicted Class A war criminal, Kishi became Japan's prime minister in 1957.

During the war, Kishi was the acknowledged leader of the reform bureaucrats, who included, among others, Hoshino Naoki, Shiina Etsusaburō, Minobe Yōji, Sakomizu Hisatsune, Mōri Hideoto, and Okumura Kiwao. The reform bureaucrats operated at the heart of the state's extensive planning apparatus that spanned Japan and Manchukuo. In Manchuria, these bureaucrats spearheaded the industrialization drive, recruited industrialists from Japan, and mediated between the Kwantung Army and Japanese mainland interests. In Japan, they drafted the government's economic control policies and plans for the New Order and promoted the official concepts of the so-called advanced national defense state and Greater East Asia Co-Prosperity Sphere.

The reform bureaucrats are the central figures in this book. Their ideas and actions reveal a new technocratic mindset and mode of power that became more explicitly defined after 1945. As technocrats, they claimed legitimacy based on their technical expertise and downplayed their ambitions for power. During the wartime period, they aimed at self-sufficiency and Asian hegemony through the creation of a technologically advanced, self-sufficient empire based on Nazi-inspired concepts of "leadership principle," "living space," and "national land planning." In the postwar era, they aimed at profits and material affluence through the establishment of a middle-class society founded on "peace and democracy."

Similar to technocrats in other industrialized countries, these leaders believed that the machine, mass production, and the modern corporation called forth the need for state planning and managerial expertise. They argued that the classical liberal ideas of "economic man," private profits, and laissez-faire were being replaced by the new technocratic principles of "organization man," public profits, and planning. Of the three types of planning—democratic, fascist, and socialist—they believed that fascist planning was best suited for ambitious "have-not" countries such as Japan, Germany, and Italy. I examine their technological worldview, occupational interests, and political alliances and how they are linked to fascism. How did these idealistic "officials of the emperor" come to join hands with the military, the right, and the left in the 1930s? What drove them to pursue a fascist route to power through imperialism, war, and a total reordering of state and society?

Reform bureaucrats promoted a radical, authoritarian form of technocracy, which I refer to as "techno-fascism." They sought to realize a productive, hierarchical, organic, national community based on the cultural and geopolitical notions of Japanese ethnic superiority and the managerial principles of "fusing private and public" and "separating capital and management." As a mode of politics, techno-fascism represented a new form of authoritarian rule in which the "totalist" state is fused with the military and bureaucratic planning agencies and controlled by technocrats.[3] It signified neither the rise of militarism nor the reassertion of traditional military-bureaucratic rule, but rather the ascendance of a new group of technocratic leaders who operated at the heart of the wartime system. Like other Japanese fascist visions and programs advanced by radical young officers and right-wing activists, techno-fascism embraced narrow, authoritarian rule; a form of state-directed economy; and an ethnic chauvinist, community-centered ideology that was used to justify war and imperialist expansion. In contrast to other fascist visions and programs, however, techno-fascism appealed to professionals on the left and right. It deflected sensitive distributional issues about winners versus losers, costs versus benefits, and ends versus means by promising increased productivity, efficiency, and co-prosperity through superior technology, organization, and national spirit.

Both the Japanese and English language scholarship have primarily analyzed Japanese fascism within an interpretive framework that emphasizes Japan's lack of modernity. Thus Japanese fascism is commonly identified with the romantic, premodern, irrational views of the radical right associated with Japanese agrarianism, ultranationalism, Japanism, and pan-Asianism. Its key representatives have been identified as right-wing activists such as Kita Ikki, Ōkawa Shūmei, and Nakano Seigō. Like the notion of a German *Sonderweg*, Japan's fascist route has been viewed as a deviation from the traditional Western path of modernization culminating in liberal capitalist democracy. According to this view, Japan's modernization was incomplete because of its late development, weak bourgeois democracy, and authoritarian, backward-looking leaders. The Japanese historian Maruyama Masao argued that fascism originated in a reactionary movement from below and later took the form of a military bureaucratic dictatorship from above.[4] He believed that a popular fascist movement was not realized because Japanese politics lacked deep democratic roots and a mass base. At the same time, Japanese fascist rule "from above" was "dwarfish." In contrast to strong, charismatic leaders like Hitler or Mussolini, Japanese leaders were "pathetic robots, manipulated by outlaws."[5] For Maruyama, Japanese fascism both originated in and was limited by the country's premodern, feudal character.

---

3. See the definition of "techno-bureaucratic fascism" advanced by Georges Gurvitch in *The Social Frameworks of Knowledge* (New York: Harper & Row, 1972), 208.

4. Maruyama Masao, "The Ideology and Dynamics of Japanese Fascism," *Thought and Behaviour in Modern Japanese Politics* (London: Oxford University Press, 1966).

5. Ibid., 92.

Those who reject the fascist label for Japan adopt a narrow definition of fascism based on the historical experiences of Fascist Italy and Nazi Germany. They view the absence of a mass party and charismatic leader—or absence of "mass modernity"—as sufficient grounds for abandoning the term altogether.[6] This view fails to give adequate weight to fascism's ideology, expansionist policies, and modern technocratic thrust—all of which Japanese fascism shared with its European counterparts. Moreover, it fails to capture fascism's politically innovative technique of transcending the left and right by appealing to both technology *and* culturally specific notions of a master people.[7]

This book contributes to a growing body of scholarship that reassesses the problem of Japanese fascism and fascism more generally.[8] I argue that in Japan, fascism was not the preserve of the zealous, irrational right wing. Wartime planning documents, essays, speeches, and interviews with Japanese technocrats suggest that a wide range of professionals embraced its ideology, policies, and programs. Fascism offered a means to overcome the crisis of capitalism and resolve the problems of class conflict and authority in modern industrial society. It was viewed as a "third way," an alternative path to modernity that was superior to liberalism and communism and best suited to meet the technological challenges of the modern era. Two convictions lay at the core of this techno-fascist vision. First, Japanese technocrats believed that "have-not" countries such as Japan, Germany, and Italy could overcome their resource deficiencies through technology and national spirit. Second, Japan was destined to become the ruler of Asia. In the fascist worldview, the globe would be reorganized geopolitically into four self-sufficient "pan-regions" ruled by the United States, the Soviet Union, Japan, and Germany.

---

6. See, for instance, Stanley Payne, "Fascism, Nazism, and Japanism," *International History Review* 6, no. 2 (May 1984); Roger Griffin, *The Nature of Fascism* (London: Routledge, 1991); and Robert O. Paxton, *The Anatomy of Fascism* (New York: Alfred A. Knopf, 2004), 197–204. Works which offer a nuanced treatment of Japanese fascism but ultimately reject the term include Peter Duus and Daniel Okimoto, "Fascism and the History of Pre-War Japan: The Failure of a Concept," *Journal of Asian Studies* 39, no. 1, 65–76; Gregory Kasza, "Fascism from Above: Japan's *Kakushin* Right in Comparative Perspective," in *Fascism outside Europe*, Stein Ugelvik Larsen, ed. (Boulder, CO: Social Science Monographs, 2001).

7. See Götz Aly and Susanne Heim, *Architects of Annihilation* (London: Weidenfeld and Nicholson, 2002); Jeffrey Herf, *Reactionary Modernism: Technology, Culture, and Politics in Weimar and the Third Reich* (Cambridge: Cambridge University Press, 1986); Zeev Sternhell, *Neither Right nor Left: Fascist Ideology in France*, trans. David Maisel (Princeton: Princeton University Press, 1986).

8. Andrew Gordon, *Labor and Imperial Democracy in Prewar Japan* (Berkeley: University of California Press, 1991); Yasushi Yamanouchi, J. Victor Koschmann, and Ryūichi Narita, eds., *Total War and "Modernization"* (Ithaca: Cornell East Asia Series, 1998); Harry Harootunian, *Overcome by Modernity* (Princeton: Princeton University Press, 2000); E. Bruce Reynolds, *Japan in the Fascist Era* (New York: Palgrave, 2004); Alan Tansman, ed., *The Culture of Japanese Fascism* (Durham, NC: Duke University Press, 2009).

I analyze Japanese techno-fascism as a product of both long-term systemic developments and historical ruptures. Techno-fascism was made possible by the technological developments of the late nineteenth and early twentieth centuries and the emergence of a professional class. At the same time, major historical events created opportunities for Japanese leaders to radically change traditional approaches to industry, war, and empire, and form new alliances. Chapter 1 examines the ways in which technological developments and the transformative events of World War I and the Great Depression produced a new type of leader in Japan's military, industry, and bureaucracy.

The Manchurian occupation was a pivotal event in the rise of Japanese techno-fascism. Chapter 2 shows how Manchuria represented an experimental ground for military fascism in the early 1930s. In Chapter 3, I analyze the process by which reform bureaucrats altered the military's program and strategies of Manchurian development and promoted their own technocratic agenda for Manchukuo. The experience of industrial development and state-building in Manchuria profoundly changed the political orientation of these bureaucrats. On returning to Japan, they formulated their own techno-fascist visions for Japan and Asia. Chapter 4 focuses on the ideas of Okumura Kiwao and Mōri Hideoto, the acknowledged ideologues of the reform movement. Chapter 5 examines the attempts of reform bureaucrats to establish a domestic New Order and the political resistance they faced from Japan's conservative establishment.

War and defeat laid the grounds for the transition from techno-fascism to postwar managerialism. In Chapter 6, I show how the Pacific War provided a timely opportunity for the reform bureaucrats to complete the unfinished business of the New Order and try to change Japan's bureaucratic ethos and culture. I examine their attempts to portray the Pacific War as an ideological war to "liberate Asia." Japan's defeat and occupation by the United States in August 1945 represented another opportunity for a major remaking of the Japanese state. In the Epilogue, I consider the favorable historical conditions for the ascendance of Japanese technocrats and their legacy in the creation of Japan's postwar democratic system.

JAPAN'S WARTIME TECHNOCRATS

In his theory of what he termed the "managerial revolution," James Burnham proclaimed that capitalism was coming to an end. What was emerging in its place was not socialism but a new type of "managerial" society.

> What is occurring in this transition is a drive for social dominance, for power and privilege, for the position of ruling class, by the social group or class of the *managers*.... This drive, moreover, is world-wide in extent, already well advanced in all nations, though at different levels of development in different nations.[1]

Burnham, a former adherent of Trotskyism, rejected Marx's theory of class struggle, by which capitalists would be overthrown by workers. Contrary to the Marxist historical view, he argued, the capitalist class was being replaced, not by the proletariat, but by a new quasi-class of "managers." These managers did not own the means of production but controlled them through their posts within the state bureaucracy. Moreover, under their leadership, societies were becoming increasingly totalitarian. Burnham believed that the Soviet Union and Nazi Germany had advanced the farthest along this route, while New Deal America was beginning to reveal similar tendencies.

Burnham articulated a vision of the modern world that was shared by many technocrats and their theorists in the advanced industrial countries. Japanese

---

1. James Burnham, *The Managerial Revolution: What is Happening in the World* (Bloomington, IN: Greenwood Press, 1941), 71–72.

technocrats, like their counterparts in the West, perceived their country to be part of the global trend of technocratic modernity. Its institutional symbols—Nazism, Stalinism, and New Dealism—suggested that the developmental paths of industrial countries were converging toward centrally planned, professionally administered mass societies. According to this view, the application of the principles of the machine to industry and government was bringing about the increased organization of work and "technicization" of society. Laissez-faire capitalism, in which individual capitalists maximize profits in response to market supply and demand, was being replaced by organized capitalism, which was technology-driven, hierarchical, and centered on the large organization. Eventually, they believed, capitalist society would give way to a new form of managerial society in which the ownership and control of capital were separate and political power was transferred to a new professional class of technologically oriented managers, or technocrats.

Burnham's theory provides insights into understanding the links between technology, the formation of Japan's managerial elite, and fascism in wartime Japan. His prediction that all managerial societies inevitably become totalitarian has been disproven.[2] But he correctly pointed out how technocracy contained certain fascist tendencies. In the case of Japan, these tendencies became more pronounced as a result of the Great Depression, the Japanese occupation of Manchuria, and war mobilization. In Japan and Manchukuo, military and civilian technocrats challenged the prerogatives of Japan's traditional ruling elite and the capitalist system, through which it maintained its power and the allegiance of other social strata.[3] These technocrats undermined the principles of capitalist property relations by advocating the separation of capital and management; state controls over industry, labor, and finance; and public over private profit. Due to their indispensable managerial functions and organizational expertise, they wielded strategic power in the economy, government, and the military. They operated at the central nodes within the complex network of all large organizations. In contrast to capitalists, whose rule was indirect and mediated through the bureaucracy and political parties, technocrats as a group could exercise direct and unmediated power because their influence penetrated deeply into the institutions of modern society. If effectively coordinated and mobilized, technocrats could exercise total authoritarian power in society. During the 1930s, Japan's conservative establishment became increasingly alarmed by the subversive, anticapitalist thrust of technocratic policies and programs that threatened its

---

2. Ibid. vii–viii. In his Preface to the 1960 edition of the *Managerial Revolution,* Burnham concedes that managerial societies have exhibited a greater range of forms.

3. Alvin W. Gouldner, *The Dialectic of Ideology and Technology: The Origins, Grammar, and Future of Ideology* (New York: Seabury Press, 1976), 229–249.

power and privileges. The *zaibatsu,* in particular, fought tooth and nail against technocrats to defend the capitalist "status quo."

This chapter examines the ideas and strategies of the managerial elite in interwar Japan. During the 1930s, three groups were identified as representing a new political and economic force in Japan. The press referred to them as the "new military men," "new zaibatsu," and "new-new bureaucrats" or "reform bureaucrats." What was "new" about these groups was that they offered innovative, antiliberal approaches toward war, industry, and government that distinguished them from traditional military officers, industrialists, and bureaucrats. The new military men, associated with the army's Control faction, introduced a new scale and type of war mobilization in preparation for "total war." New zaibatsu industrialists devised a distinct management philosophy and corporate structures for heavy and chemical industries. Reform bureaucrats advocated a new activist, goal-oriented approach toward government and paved the way for unprecedented state control of politics, private industry, and public services based on their vision of the "managerial state." Within their respective areas of expertise, these professionals expressed common views and concerns about the transformative impact of technology on society and how to guide that change. They shared the conviction that the challenges arising from industrial capitalism, technological advance, and mass society called forth a new type of leader who possessed technical expertise, stood above narrow private interests, and grasped the integrated mechanisms of modern industrial society. Together they sought to reorganize Japanese society along a new ideological basis. We first consider the role of technology in the rise of this new type of society and leader.

Defining Technology and Technocrat

Technology's Impact

For the post–World War I generation of Japanese leaders, the keynote of modernity was technology. The symbols of modernity were the machine, train, car, tank, telegraph, and radio. These products of the "second industrial revolution" were made possible by the recent scientific and technological advances, and especially the widespread application of electric power to industry. By the turn of the twentieth century, Japanese technology had developed well beyond the small-scale, labor-intensive techniques of the past, although these continued to coexist with modern methods throughout the interwar period.[4] It surpassed the

---

4. Tessa Morris-Suzuki, *The Technological Transformation of Japan* (Cambridge: Cambridge University Press, 1994), 14, 35, 86.

technology of Meiji, embodied in advanced Western machinery, model plants, and foreign technicians, which was directly imported by the state. From the time of World War I, during which Japan profited immensely from favorable trade relations and new markets, it began to establish the foundations of its heavy and chemical industries.

Modernity meant not only the arrival of the machine, but also the application of its principles to society at large. Technology represented a new form of technical rationality expressed in the mass assembly line, the corporation, and the large, complex bureaucracy. It can be defined as the application of science to daily life by means of the dual processes of specialization and integration.[5] In the case of an automobile, its production required the division of labor into specialized tasks in order to apply scientific knowledge in the manufacture of its individual components, such as the engine, tire, or window. At the same time, it involved the integration of these various tasks and components to create the final product. Technocrats were engaged in the latter task as managers and administrators, whose technical grasp and organization skills enabled them to bring together the various elements and processes into a coherent whole.

Technology's potential impact could be detected most clearly in the most technically advanced industries, where the worker's role became redefined in the production process. In the mass assembly line, the worker performed specialized tasks, whose results comprised one small component of the final product. The plant, that "highly complex organization of men, machines and tools," became the productive unit or means of production.[6] The significance of mass production technology lay not in its function as a "mechanical principle" for organizing material processes, but as a "social principle" for organizing human activity to perform specific tasks.[7] As a result, mass production brought about the prioritization of technology over labor. Marx's claim that technology was a means of labor was reversed: labor now became a means of technology.[8] The organization replaced the individual worker as the unit of production and became the complement of specialization. The organization, in the form of the plant, industrial

5. For definitions of technology, see John Galbraith, *The New Industrial State* (Boston: Houghton Mifflin Company, 1967), 12–13; Herbert Marcuse, "Industrialization and Capitalism in Max Weber," *Negations: Essays in Critical Theory* (Boston: Beacon Press, 1968), 205; Peter Drucker, *The New Society* (New York: Harper & Brothers, 1950), 1–17.

6. Drucker, *The New Society*, 5.

7. Ibid., 2–6.

8. Kawahara Hiroshi develops this theme in his writings on Japanese technocracy. See "Fuashizumu e no michi," in Kawahara Hiroshi, Asanuma Kazunori, et al. eds., *Nihon no fuashizumu* (Tokyo: Yūhikaku, 1979), 16–17; and Kawahara, *Shōwa seiji shisō kenkyū* (Tokyo: Waseda daigaku shuppan-sha, 1979), 3–34.

enterprise, state trust, or government planning agency, became the institutional expression of modern technology.

With technological advance, organizations became large, costly, and complex.[9] Modern technology required an increased division of tasks and detailed coordination of a larger number and variety of highly specialized workers and equipment. Both aspects required more time for completion and larger amounts of investment capital to finance the specialists, managers, equipment, and research expenses. In order to capture the economies of scale, capital was increasingly concentrated and controlled by the corporation or the state. In effect, the individual was subsumed by the organization; management became increasingly impersonal and formalistic; and custom, tradition, and private considerations gave way to the scientific criteria of efficiency and economy.[10]

Technology, in the form of the organization, brought about a greater need for planning and an increasingly planned and organized economy. Planning can be defined as the temporal forecasting and arrangement of materials and labor in order to provide a future product or service. Planning is enhanced by the new techniques of scientific management and industrial rationalization. Taylorist methods of standardization or inventory control helped increase productivity and avoid future disruptions by establishing criteria to ensure the consistent quality or quantity of materials, labor, and goods. Industrial rationalization, through interfirm cooperation, cartels, and trusts, aimed to increase efficiency and productivity by minimizing the waste of material and effort, especially overproduction and excess competition. At the same time, rationalization gave rise to technical inefficiencies as large firms fixed prices, reduced competition, and stifled innovation in order to secure profits. As Thorstein Veblen first pointed out, capitalism became incompatible with technological progress because it subordinated it to nontechnical ends in the pursuit of profit. Technology paved the way for a more organized form of capitalism in which the economy was increasingly planned by cartels and trusts and eventually by the state.

When applied to society at large, planning became the means by which the machine refashioned the worker and society into its own image and made them compatible with its principles. Scholars have described planning as a form of "social technology" or "social technique," which introduced the new principles of machinelike efficiency and organization and brought about an advanced stage of rationalization that societies are unable to control or resist. The machine brought

---

9. Alvin Gouldner, "Metaphysical Pathos and the Theory of Bureaucracy," *American Political Science Review* 59, no. 2 (June 1955), 500–501.

10. Karl Mannheim, *Man and Society in an Age of Reconstruction* (New York: Harcourt, Brace & Co., 1954), 55.

about an increasingly planned or "technicized" society—in industry, administration, war, work, leisure, health, child-rearing, and science.[11] With regard to the latter, scientific research became increasingly planned and oriented toward technical improvement and practical application. Invention by individual, chance discovery as in the case of Edison, Bell, and Marconi, was replaced by a new military-oriented, state-planned science, or "big science," based on large research facilities and staff and incurring huge expenses and risks underwritten by the public and the state.[12] Science and technology became intertwined as technology was increasingly driven by science and science was technicized. The Japanese even formulated a new term "science-technology" (*kagaku-gijutsu*) to acknowledge this fusion.

The Technocrat's Role

Technocrats are defined foremost by their managerial function. They are responsible for the administration and coordination of the production process and strategic planning within the organization. They occupy positions in industry and government as production managers, operations directors, commissioners, bureau chiefs, and administrative engineers—among which the latter term was directly adopted by Japanese government engineers to describe their own managerial function.[13] In interwar Japan, technocrats played an integral role in mediating between the interrelated processes of administrative and technological advance. Their functions were neither purely bureaucratic nor technical. The classic definition of a technocrat—one who wields political power based on a claim to scientific or technical expertise—tends to place greater emphasis on the aspect of technical knowledge rather than political power. But the technocratic function is not that of the technical specialist as chemist, physicist, or engineer, although some began their careers in that capacity. The technician's role becomes technocratic only when they are appointed to positions of influence in which they can combine technical expertise with political power and translate planning into policy.[14]

Technocrats pride themselves on being able to see both the forest *and* the trees; their expertise, however, was not specialized but "functionally polyvalent."[15]

11. On the technicization of society, see Georges Friedmann, "Technological Change and Human Relations," *British Journal of Sociology* 3, no. 2 (June 1952), 95–116.

12. Jacques Ellul, *The Technological Society* (New York: Vintage Books, 1964), 7–11.

13. Miyamoto Takenosuke, *Tairiku no keizai kensestu* (Tokyo: Iwanami shoten, 1941), 152.

14. Magali Sarfatti Larson, "Notes on Technocracy: Some Problems of Theory, Ideology, and Power," *Berkeley Journal of Sociology* 17, no. 5 (1972–1973), 7.

15. Ibid., 11.

Their actual function is better captured by the term *techno-bureaucrat*—one who possesses strategic power and a global vision over the technically driven and administered whole.[16] Technocrats are primarily involved in the technological processes of integration, not specialization, and act in the capacity of managers and administrators of technical specialists. The term *techno-bureaucrat* distinguishes those technocrats "who are in a position to gain support from a bureaucratic machine, or manipulate its fundamental positions."[17] They are experts in not only the "administration of things," as Saint-Simon wrote, but also in the "administration of men" and include certain high-level bureaucrats and military officers, industrialists, labor union leaders, and intellectuals employed in think-tanks and planning agencies. Their career paths vary. In Japan, most new zaibatsu executives began their careers as engineers or scientists, founded their own companies, and later collaborated with the government. Others, such as the reform bureaucrats or military planners, rose within the ranks of the bureaucracy and acquired technical expertise by working in supraministerial planning bureaus, studying abroad, and managing technical projects at home or in the colonies. Their technical knowledge, together with important political connections formed through the university, the family, regional ties, and work, propelled them to leading positions in technologically strategic areas. Their authority derived from their managerial positions within the organizational structures, which privileged them with broad, multidimensional perspectives and power.

As managers rather than owners of capital, Japanese technocrats possessed different goals and interests from capitalists. In contrast to capitalists, whose performance was measured by the rate of return or profit on their investments, technocrats were judged by the technical soundness and efficient operation of the firm. Capitalists depended on the preservation of capitalist property relations and conditions for private investment and gain, including the protection of private property, a stable currency, and free access to labor, materials, energy, and foreign trade. In contrast, the technocrat was not dependent on capitalist property relations and could perform a similar function within an alternative, noncapitalist type of system.

Japanese technocrats wielded influence primarily through the strategic use of state power to back their policies and programs. Through the agents of the state, such as the military, bureaucracy, police, and courts, they challenged the prerogatives of capitalists by assuming increased control over the access to

---

16. On "techno-bureaucrats," see Larson's discussion of the term advanced by Nora Mitrani, in ibid., 11–17; Jean Meynaud, *Technocracy* (London: Faber and Faber, 1964), 65–66; Gurvitch, *The Social Frameworks of Knowledge,* 110–114, 207–212.

17. Meynaud, *Technocracy,* 65.

and distribution of the means of production. The Japanese Army, in particular, through military invasion and total war mobilization, provided opportunities for a radical reorganization of Japanese society and the ascendance of technocrats to power. The industrialization and globalization of warfare, which characterized the Russo-Japanese War and World War I, required the total mobilization of society for war and long-term planning. Military officers, together with technical experts, economic bureaucrats, and industrialists, drafted a series of economic control laws, production expansion plans, and the National General Mobilization Law that empowered the state to intervene in the private sphere.

Wartime technocrats extended their political influence by gradually infiltrating the bureaucracy and usurping the planning functions of the nontechnical ministries. From 1927, they established a series of supraministerial economic bureaus and planning agencies, which by the early 1940s, comprised a technical substructure within the Japanese bureaucracy. Within this planning apparatus, which spanned Japan and Manchukuo, military and civilian technocrats drafted the major economic control legislation for Japan and its empire. These wartime emergency laws enabled them to assume effective control over capital within an increasingly controlled economy.

The diverse backgrounds and eclectic approaches of technocrats enabled them to draw support from both the left and right in denouncing capitalists and their supporters. Technocrats approached the problem of planning from a variety of viewpoints. Those tending toward conservative or right-wing views, many of whom were trained in German law or military strategy, saw planning as a means to enhance the power and authority of the state over private interests. Progressive left-wing technocrats, who in their youth had flirted with Marxism and envisioned themselves as the intellectual leaders of an incipient workers' movement, looked to planning to reduce social tensions, increase industrial efficiency and production, and improve the conditions of labor. Their technocratic vision affirmed neither capitalism nor socialism, but managerialism. Moreover, their call for expertise as the primary criteria for leadership, over wealth, peerage, or official rank, appealed to middle-class professionals, such as engineers, scientists, political activists, journalists, teachers, and intellectuals, who felt excluded from the privileges of power. Like other members of the intellectual elite, technocrats saw themselves as interpreters of the latest Western theories and political trends.[18]

---

18. On Japanese intellectuals and fascism, see William Miles Fletcher, *The Search for a New Order: Intellectuals and Fascism in Prewar Japan* (Chapel Hill: University of North Carolina Press, 1982).

Control Officers and Total War

The Japanese Imperial Army has been portrayed as the bastion of feudal reaction, irrational militarism, and emperor worship.[19] Certainly from its creation in the 1870s, the army had been a traditional foe of party politics and laissez-faire capitalism. The oligarch Yamagata Aritomo designed the Japanese Army along the lines of the German Imperial Army. He envisioned it as a central force in a Prussian-style military-bureaucratic government centered on the emperor and dominated by leaders from Chōshū. This static image of Japanese militarism, however, fails to reflect the ways in which the army responded to new political trends, technological advance, and two major global wars.[20] From the time of the "Taishō Political Crisis" in 1912, party politicians challenged the army's strong-arm tactics. During the 1920s, the military's influence and status diminished as Japanese leaders embraced party cabinets, liberal capitalism, and cooperative diplomacy, which included naval arms limitations and caps on army spending. More important, in response to the lessons of the Russo-Japanese War and World War I, a new professional class of ascendant army officers criticized not only "money politics" but also the army's traditional region-based cliques and emphasis on spiritual mobilization. These officers advocated troop reductions and mechanization in order to make the army leaner and more effective. Drawing on the lessons of the recent wars, they promoted modern technocratic visions of "total war" and a "national defense state."

The new generation of army technocrats was affiliated with the "Control faction" (Tōsei-ha) under Nagata Tetsuzan. The Control faction derived its name from its strong opposition to the insubordination of radical officers in the early 1930s and call for more "control" within army ranks. Prominent members included Ishiwara Kanji, Tōjō Hideki, Suzuki Teiichi, Akinaga Tsukizō, Ikeda Sumihisa, Katakura Tadashi, Tanaka Kiyoshi, and Mutō Akira. The Control faction was distinct from other army groups such as the Imperial Way faction (Kōdō-ha), radical young officers, and Purification Group faction (Seigun-ha).[21] Whereas the latter groups emphasized indoctrination in the principles of the imperial way, warrior spirit, and loyalty to the imperial polity (kokutai), the Control faction focused on technological advance, army mechanization, and economic

---

19. The classic account in English is Ienaga Saburō, *The Pacific War, 1931–1941* (New York: Pantheon Books, 1978), particularly chapter 3.

20. John W. Steinberg, Bruce W. Menning et al. eds., *The Russo-Japanese War in Global Perspective: World War Zero* (Leiden: Brill, 2005); Michael Barnhart, *Japan Prepares for Total War* (Ithaca: Cornell University Press, 1987); Alvin D. Coox, *Nomonhan: Japan Against Russia, 1939* (Stanford: Stanford University Press, 1985).

21. James B. Crowley, "Japanese Army Factionalism," *Journal of Asian Studies* 21, no. 3 (May 1962), 309–326.

planning. From the late 1920s, they formed study groups such as the Futabakai, Issekikai, and Mokuyōbikai to discuss a wide range of issues including upgrading the army's equipment, eliminating regional cliques, mass mobilization, and resolution of the "Manchurian problem."

Their social background, training, and intellectual orientation distinguished Control officers from the rank-and-file. Lower-ranked officers were predominantly the sons of small landowners, independent cultivators, and shopkeepers, often from the less-prosperous northeastern regions of Japan. The basic military and ideological training they received at the Military Academy prepared them for battle and instilled in them a strong sense of warrior spirit and loyalty to the emperor. Through participation in clandestine patriotic societies such as the Imperial Sword Party, they became acquainted with the writings of right-wing thinkers such as Kita Ikki, Ōkawa Shūmei, Gondō Seikyō, and Tachibana Kōsaburō.

Control officers formed a part of Japan's intellectual elite. Most were middle- or upper-middle-class graduates of the highly competitive War College. Like their civilian counterparts within the bureaucracy, they were the products of an emerging meritocratic system that increasingly rewarded ability over social status and regional ties. Similar to Japanese intellectuals, these officers were keen to absorb the latest Western theories and ideas such as those of Ludendorff, Clausewitz, Delbrück, and Haushofer. Many had rounded out their military education with additional study at elite universities in Japan and abroad. More than other high-ranking officers, they demonstrated an aptitude for theory, which they further developed by studying Marxism. In his youth Suzuki Teiichi was influenced by the writings of the Marxist economist Kawakami Hajime, particularly his *Tale of Poverty (Bimbō monogatari)*. After completing his military education, Suzuki studied economics for a year and briefly worked at the Ministry of Finance.[22] Both Akinaga and Ikeda rounded off their military training with three years of study at the economics faculty of Tokyo Imperial University, where they immersed themselves in the debates on capitalism and Marxism. In Manchuria these officers subsequently honed their skills in the elite planning unit of Japan's Kwantung Army.

Their vision of modern warfare centered on the concept of total war derived from their firsthand observations of World War I.[23] The war signified the first

---

22. Suzuki Teiichi, *Suzuki Teiichishi danwa sokkiroku* (Tokyo: Nihon kindai shiryō kenkyūkai, 1971), 189.

23. See Mikuriya Takashi, "Kokusaku sōgō kikan setsuoku mondai no shiteki tenkai: Kikakuin sōsetsu ni itaru seiji ryoku gaku," in Kindai Nihon kenkyūkai-hen, *Nenpō kindai Nihon kenkyū*, vol. 1, *Shōwaki no gunbu* (Tokyo: Yamakawa shuppansha, 1979), 125.

"state total war": extended wars conducted between nations, not just troops, requiring the mobilization of the entire population and self-sufficiency in resources. Future wars would be "total" because they are fought on a number of fronts, command the entire resources of the nation, and potentially bring about mass destruction on a global scale.[24] Japan's total war strategist Ishiwara Kanji advanced an apocalyptic vision of a "final war" or a "war to end all wars."[25] Although the Russo-Japanese War introduced a new type of industrialized, global warfare, World War I signified the first war of attrition requiring a new scale of national mobilization.[26]

In the fall of 1915, the Japanese Army commissioned a study of the doctrines and mobilization policies of the European countries at war. The study highlighted their shortages in materiel and troops and the need for more comprehensive resource mobilization.[27] In 1917, Colonel Koiso Kuniaki produced a top secret, but highly influential, report entitled *Resources for the Defense of the Empire* (*Teikoku kokubō shigen*) that laid out the requirements for total war and introduced the concepts of "national general mobilization" and "economic self-sufficiency." Koiso envisioned a new type of "national defense economy" to promote exports, curtail imports, preserve domestic resources, stockpile war materials, convert civilian industries to munitions production, and maintain close contact with the Asian continent.[28] After directly witnessing World War I as an official observer, Nagata and two army colleagues, Okamura Yasuji and Obata Toshirō, met at a hot springs resort in Baden-Baden, Germany, in late October 1921 and mapped out Japan's future defense strategy.[29] After Nagata returned to Japan, he pushed for reforms to mechanize and rationalize the army and develop the nation's resources for total war. These plans formed the basis for Japan's Munitions Law of 1918.

24. Bōeichō bōei kenshūjo, *Rikugun gunju dōin, vol. 1, Keikaku-hen* (Tokyo: Asagumo shinbunsha, 1967), particularly chapter 2; Hans Speier, "Ludendorff: The German Concept of Total War," in Edward Mead Earle, ed., *Makers of Modern Strategy: Military Thought from Machiavelli to Hitler* (New York: Atheneum, 1966).

25. On Ishiwara see Mark R. Peattie, *Ishiwara Kanji and Japan's Confrontation with the West* (Princeton: Princeton University Press, 1975).

26. Recently scholars have argued that the Russo-Japanese War represented the first "world war" by introducing a new type of industrialized, technological, and global warfare. From the perspective of Japanese army technocrats, however, the total war concepts of autarky and national general mobilization were first introduced into planning as a result of World War I. See Steinberg, Menning, et al. eds., *The Russo-Japanese War in Global Perspective*.

27. For an analysis of these reports, see Bōeichō bōei kenshūjo, *Rikugun gunju dōin, vol. 1*, 14–32.

28. Ibid., 41.

29. Mikuriya, "Kokusaku sōgō kikan setsuoku mondai no shiteki tenkai," 125.

Technological innovation and the rise of mass society shaped the total war vision. It drew on the latest technical advances in the heavy and chemical industries, communications, and social management to produce sophisticated weapons, develop natural and synthetic resources, and mobilize the masses. Already during the war, the Japanese Army and Navy began to step up efforts to improve their technical knowledge base in such areas as aircraft, synthetic petroleum, and optics by establishing its own research centers and recruiting scientists and engineers from Japan's leading universities.[30] Manchuria, which had come under Japan's indirect control as a result of victory in the Russo-Japanese War, was targeted as a resource base for heavy industry in the army's drive for autarky. World War I also encouraged a new way of thinking about the economy and society and how to effectively mobilize and allocate the nation's human and material resources. General Erich Ludendorff's *Total War (Der totale Krieg)* had argued that the underlying reason for Germany's defeat was not so much its inadequacies in military strength or strategy, but rather its failure to prepare for war in other areas such as economic planning and mass propaganda. As the Japanese Army concluded from the German experience: "Their downfall, in short, was that they could not endure the economic blockade of the great powers, the people lacked adequate nutrition, the energy to resist was diminished, and furthermore, as a result of the propaganda wars the people lost their fighting spirit and revolutionary ideas came to the fore; the unavoidable consequence was an internal breakdown and hasty solicitations for peace."[31] In order to sustain popular morale during a long war, modern governments required not only effective technical means to communicate to the masses but also strategies to organize the civilian war effort. For the above reasons, army planners characterized modern wars as "battles of organizational ability":

> The form of war as an international struggle for survival is becoming scientific and organizational, especially with respect to the propaganda wars, economic wars, and military wars. To put it in the extreme, we can say that future international disputes will be a competition between intellect and intellect, a struggle of organization against organization. Therefore, we can probably say that the honor of victory will be given by the opponent to the one who possesses superior *creativity* and *organization*.[32]

30. Hiroshige Tetsu, *Kagaku no shakaishi—kindai Nihon no kagaku taisei* (Tokyo: Chūō kōronsha, 1973), 86–87.

31. "Kokubō no hongi to sono kyōka no teishō," in Takahashi Masae, ed., *Gendaishi shiryō*, vol. 5, *Kokkashugi undō* (Tokyo: Misuzu shobō, 1964), 269.

32. Ibid., 277.

As twentieth-century wars were increasingly organized along the principles of modern industry, they required greater planning and organization. The application of advanced technology required immense capital commitments and longer time frames—Ishiwara predicted that Japan would need twenty years to prepare for the final war. On the other hand, as scientific activity became technicized and monopolized by the state, as in the development of synthetic resources, it required immense amounts of funding, a research staff, and laboratories. Technocrats believed that scientific discovery and creativity could be fostered more efficiently within a state planned system rather than under market capitalism.

Following the outbreak of World War I, the army began to lay the institutional and legal groundwork for a national defense state. In 1915, the army created the Temporary Military Research Committee to conduct research on state mobilization. In the spring of 1918, it created the Munitions Industry Mobilization Law and Munitions Bureau. The Munitions Law was drafted under the Terauchi cabinet by Major Suzumura Kōichi based on his study of industrial mobilization policies in Europe. It empowered the Munitions Bureau to assess the nation's capabilities in munitions production in peacetime and to mobilize private factories for war production and even directly manage private enterprises in times of war. Originally, the army intended to grant the state greater powers of intervention and directly challenge the principles of a market economy. According to one plan advanced by the Army General Staff, the state should not only be permitted to expropriate civilian goods; manage or use factories or firms producing those goods; and order the transfer of materials, personnel, or the ownership rights of factories and firms. It should also be permitted to order a firm to divulge its proprietary designs and patents and force a firm to hand over to the state any "excess profits" made through munitions-related production.[33] These provisions were to be applied not only in wartime but in peacetime, since a "state of war" originally was interpreted to include "incidents" as well as the postwar period.[34] Although the basic legal framework was put into place early on, the military was slow to implement the law and the state's broad economic powers were not exercised until the outbreak of the China War in 1937.[35] The law formed the basis of the National General Mobilization Law of 1938, which gave technocrats a virtual carte blanche to intervene in the private sphere to mobilize the nation for war.

---

33. General Staff Chief Uehara Yūsaku submitted the plan "Gunjuhin kanri hō" to Army Minister Ōshima Kenichi on December 21, 1917. See Homma Shigeki, "Senji keizai hō no kenkyū," *Shakai kagaku kenkyū* 25, no. 6 (March 1974), 28–30; Bōeichō bōei kenshūjo, *Rikugun gunju dōin*, vol. 1, 53–54.

34. Homma, "Senji keizai hō no kenkyū," 29.

35. Hashikawa Bunzō, "Kokubō kokka no rinen," in Hashikawa Bunzō and Matsumoto Sannosuke, eds., *Kindai Nihon seiji shisōshi*, vol. 2 (Tokyo: Yūhikaku, 1970), 233.

To enhance the state's planning capacity, the army created the first of a series of cabinet-level planning organs. Nagata Tetsuzan originally envisioned a powerful national defense board to be placed directly under the prime minister. With the backing of Army Minister Ugaki Kazushige, he established the Equipment Bureau in 1925 to provision the nation for war and assumed the post of head of its mobilization section. Two years later, Nagata came closer to realizing his vision with the creation of the Resources Bureau. The new bureau took charge of all aspects of military and civilian resource mobilization and emphasized research and long-term resource development. In contrast to the Munitions and Equipment Bureaus, the Resources Bureau was primarily a civilian organ staffed by elite bureaucrats. Matsui Haruo, from the Legislative Bureau, served as head of both general affairs and planning, while Commerce Ministry bureaucrat Uemura Kōgorō became head of research. Army and navy officers on active duty were taken on as special appointees. In the new supraministerial planning agencies, members concurrently held positions within these agencies and their own ministries. The cabinet-level agencies reflected the attempt to promote unified planning at the top and overcome the problem of bureaucratic sectionalism. In order to enhance the stature and effectiveness of the new bureau, the prime minister was granted the authority via imperial command to override individual ministries in matters concerning mobilization. Technocratic planning was further strengthened by the creation of the Cabinet Planning Agency in 1935 and its subsequent merger with the Resources Bureau into the powerful Cabinet Planning Board in 1937.

The technical subsystem emerging within the bureaucracy provided technocrats with an autonomous, institutional base from which to exercise power and establish a national agenda for total war mobilization. The supraministerial planning organs set a precedent for policymaking by which economic planning was undertaken almost exclusively by technocrats and then forced on the Diet for approval in the form of mobilization laws. In passing both the National General Mobilization Law and the New Order legislation, technocrats circumvented the Diet and the political parties; greatly expanded their scope of technical administration; institutionalized access to information; and used their power to pressure the Diet to ratify their policies.[36] As a result, the Diet lost its already highly circumscribed role as a political pulpit to analyze and criticize government policy. The political parties, sensing the demise of limited political pluralism, dissolved themselves and sought to carve out a role within the new system.

---

36. Homma, "Senji keizai hō no kenkyū," 49.

Moreover, through this technical subsystem, control officers skillfully set the terms for national debate. Their vision of total war acquired near hegemonic control over wartime political discourse because of their success in defining the ways in which people thought about the war. Officers tapped into the anxieties of the new era. They presented technological developments as unquestioned historical law. Their vision appeared "scientific" and rational—a kind of social Darwinian law of the survival of the fittest—and effectively tapped into the fears of Japan's elite that their country might fail to share the world stage with the great powers. Although many bureaucrats, politicians, and business leaders denounced technocrats for their high-handedness and arrogance, they seldom questioned the legitimacy of their total war vision or Japan's need for self-sufficiency.

New Zaibatsu and Technology-based Industry

Studies of Japan's war economy have focused predominantly on the central role of the zaibatsu. In both the reports of the Supreme Commander of the Allied Powers, or SCAP, and postwar writings of Marxists, the zaibatsu were portrayed as key perpetrators and beneficiaries of the Pacific War. As the government's primary munitions suppliers from 1941 until 1945, these industrial and banking conglomerates came to dominate Japan's economy. By the end of the war, the Big Four zaibatsu, Mitsui, Mitsubishi, Sumitomo, and Yasuda, controlled nearly 25 percent of the country's total capital and 80 percent of total foreign investments. After the war the two largest zaibatsu, Mitsui and Mitsubishi, with their vast network of 250 to 300 subsidiaries, became the first targets of SCAP's program to break up the zaibatsu.

Surprisingly little attention, if any, has been paid to the new zaibatsu. As contemporary leaders readily acknowledged, the new zaibatsu were the military's most trusted allies and chief collaborators in the industrial development of Japan's empire. In contrast to the traditional zaibatsu, which throughout the 1930s staunchly defended the capitalist status quo and faced the wrath of the right, the new zaibatsu aligned themselves with the forces of "reform" and supported the drive for a national defense state, controlled economy, and autarky. At the behest of the army in 1937, Nissan founder Ayukawa Yoshisuke transferred his entire operations to Manchuria to oversee the development of heavy industry. The Nitchitsu zaibatsu directed chemical industries in Korea. One of the main reasons why SCAP downplayed the role of the new zaibatsu is that during the Pacific War their fortunes rapidly took a turn for the worse. By the end of the war, the share of Nissan's foreign investments of its total investments had fallen from 75 percent to 6 percent, while the share of overseas investments of Mitsui and

Mitsubishi had increased to 60 percent of its total investments.[37] As a result of the reversal of Nissan's fortunes, as well as Ayukawa's loss of faith in the military and belated peace maneuvers to the United States, Nissan appeared less menacing and more "liberal" than the old-line zaibatsu.[38]

The tendency to downplay or ignore the differences between the two groups also reflected the political agenda of researchers. Occupation authorities and left-wing scholars believed that all zaibatsu were fundamentally undemocratic and feudalistic. They viewed their hierarchical structure as an expression of Japan's patriarchal family system, which denied individual initiative and oppressed workers. Through their oligopolistic control of finance and a wide range of industries, the zaibatsu limited competition, manipulated prices, and squeezed out smaller firms. They used their holding companies to leverage their capital by investing in a majority share of stocks in many firms. Moreover, the zaibatsu exerted their enormous financial power in the government through their high-level connections within the bureaucracy and financial patronage of the political parties. SCAP believed that the extreme concentration of capital in the hands of a small number of families perpetuated the economy's dual structure, in which traditional, backward industries coexisted with modern, advanced ones. Occupation authorities, on the one hand, blamed the zaibatsu for hindering the growth of a healthy, liberal capitalist economy characterized by a vigorous labor market, improved working conditions, and strong consumer demand. On the other hand, Marxists believed that the zaibatsu signified the final stage of state monopoly capitalism and imperialism leading to its eventual overthrow. In sum, what mattered to these researchers was the size and extent of market power of these zaibatsu, and not the distinctions between "new" and "old." Hence they focused on the Big Four and not on the smaller upstarts.

One cannot grasp the true character of Japanese wartime leadership and ideology without understanding the distinct identity of the new zaibatsu. Technologically minded entrepreneurs founded these combines in order to take advantage of the unprecedented opportunities presented by World War I. Following the outbreak of war, multilateral trade relations were suddenly altered in Japan's favor.[39] As Europe became embroiled in war and shipping routes were blocked, its export capability rapidly declined and its vast overseas export markets became open to Japanese goods. Europe itself became a net importer of

---

37. Eleanor M. Hadley, *Antitrust in Japan* (Princeton: Princeton University Press, 1970), 41.

38. Iguchi, *Unfinished Business,* 138–170.

39. On the impact of World War I on Japan, see Hashimoto Jurō, "Kyōdai sangyō no kōryū," in Nakamura Takafusa and Odaka Kōnosuke, eds., *Nihon keizaishi,* vol. 6, *Nijū kōzō* (Tokyo: Iwanami shoten, 1989), 82–99.

munitions-related goods from countries such as Japan and the United States, which also increased its imports from Japan and other countries in East Asia. Japan's shipping industry profited enormously from both shipbuilding and soaring freight rates. For new zaibatsu leaders, however, the most important benefit of the war was the curtailment of imports of foreign heavy industry and chemical industrial products such as iron, steel, and synthetic fertilizer. Strong domestic civilian and military demand for heavy and chemical goods, coupled with abundant and relatively inexpensive electric power, provided the ideal setting to engage in heavy industrial catch-up.[40]

Those entrepreneurs who possessed a keen appreciation for the new technological advances and took advantage of the favorable economic setting became the captains of the so-called "new" or "newly emerging" zaibatsu (*shinkō zaibatsu*).[41] The new zaibatsu referred to the five large conglomerates: Ayukawa Yoshisuke's Nissan, Mori Nobuteru's Mori Concern, Noguchi Jun's Nitchitsu (Japan Nitrogenous Fertilizer Company), Nakano Yūrei's Nissō (Japan Soda), and Ōkōchi Masatoshi's Riken (Physical and Chemical Research Institute). In contrast to the Big Four zaibatsu and "Taishō zaibatsu" such as Suzuki, Kuhara, Nomura, Iwai, and Matsukata, which concentrated on mining, finance, and commerce, the new zaibatsu focused predominantly on the heavy and chemical industrial sectors.[42] After the war, they skillfully capitalized on the postwar economic recession to expand their own businesses into sprawling conglomerates.

One of the basic differences between the new and old zaibatsu was their business vision. The new zaibatsu were driven by science and technology, in contrast to the traditional zaibatsu, whose primary goal was to preserve family wealth and honor. These different goals were reflected in their corporate structures. The traditional zaibatsu possessed numerous subsidiaries in diverse, unrelated markets such as banking, insurance, coal mining, paper, machine tools, and shipbuilding. They entered these businesses as a result of special government contracts, personal connections, and market opportunity. The new zaibatsu, on the other hand, organized their various businesses around a key technology and resource such as electro-chemical technology and hydroelectric power. Their multilateral

40. Nakagawa Keiichirō, "Business Strategy and Industrial Structure in Pre-World War II Japan," in Nakagawa Keiichirō, ed., *The International Conference on Business History*, vol. 1, *Strategy and Structure of Big Business* (Tokyo: University of Tokyo Press, 1976), 26.

41. I adopt the standard English translation of "new zaibatsu" instead of the more precise translation "newly emerging zaibatsu," even though this term might cause some confusion with the group of "Taishō" zaibatsu, which some Japanese scholars sometimes refer to as "new zaibatsu."

42. Researchers agree on the classification of the Big Four zaibatsu, but not on the broader Japanese classification of "old zaibatsu" and "Taishō zaibatsu."

business structures were designed to operate in a more systematized, integrated, and organic fashion.

Most new zaibatsu specialized in the chemical industries. Noguchi founded Nitchitsu in 1908 to produce calcic nitrogen fertilizer utilizing the excess power created by his Kyūshū firm Sogi Electric.[43] His firm grew rapidly during World War I, when foreign imports of fertilizer were curtailed. In 1922 Noguchi used these profits to invest in a new plant to produce synthetic ammonium sulphate. From the mid-1920s, Nitchitsu expanded its operations in Japan-occupied Korea to take advantage of its rich resources, particularly hydro-electric power. Noguchi built an electric power facility at the Pujon River, which was the first of several electric power plants that he built in Korea, as well as additional electric power plants in Japan. He also diversified into ammonia-related products such as explosives and rayon in both Japan and Korea. Similarly, Mori and Nakano built their businesses around electro-chemical technology. After securing his own supply of electric power, Mori began production of calcium cynamide and synthetic ammonia through his Shōwa Fertilizer Company. In the 1930s he diversified into aluminum, electric copper, soda, and explosives and created Shōwa Electric.[44] Nakano launched his business career by taking over a zinc refining firm and subsequently merging it under the newly formed Nissō. During the war Nissō produced electrolysis soda and later expanded into mining, iron, steel, rayon, and pulp.[45] Ōkōchi founded the Riken Company (Riken kōgyō) in 1927 to commercialize the scientific and technological inventions of the semipublic Riken and to support future research. Based on its domestic and foreign patents, Riken established subsidiaries to commercialize its findings such as Riken Magnesium, Riken Piston Ring, Riken Electric Wire, and Riken Corundum, and by 1937 had established thirty-one subsidiaries in Japan.[46]

---

43. Barbara Molony, "Innovation and Business Strategy in the Prewar Chemical Industry," in Yui Tsunehiko and Nakagawa Heiichirō, eds., *The International Conference on Business History*, vol. 15, *Japanese Management in Historical Perspective* (Tokyo: University of Tokyo Press, 1989), 24; Barbara Molony, *Technology and Investment: The Prewar Japanese Chemical Industry* (Cambridge: Harvard University Asia Center, 1990).

44. Mikami Atsufumi, "Old and New *Zaibatsu* in the History of Japan's Chemical Industry: With Special Reference to the Sumitomo Chemical Co. and the Shōwa Denko Co.," in Ōkochi Akio, and Uchida Hoshimi, eds., *The International Conference on Business History*, vol. 6, *Development and Diffusion of Technology: Electrical and Chemical Industries* (Tokyo: University of Tokyo Press, 1980), 216.

45. Udagawa Masaru, *Shinkō zaibatsu* (Tokyo: Nihon keizai shinbunsha, 1984); Arisawa Hiromi, ed., *Shōwa keizai shi* (Tokyo: Nihon keizai shinbunsha, 1976), 99–102.

46. Udagawa, *Shinkō zaibatsu*, 202–206; Samuel K. Coleman, "Riken from 1945 to 1948: The Reorganization of Japan's Physical and Chemical Research Institute under the American Occupation," *Technology and Culture* 31, no. 2 (April 1990), 229–231.

The different capital structures of the old and new zaibatsu reflected their different business strategies. The former were owned and tightly controlled by the founding family and characterized by their "closed finance."[47] Mitsui, Mitsubishi, Sumitomo, and Yasuda possessed their own bank, which funded most of the activities of their various subsidiaries. When these firms incorporated some of their businesses, they did so not to develop new technology-based industries but to secure majority control over their enterprises. The new zaibatsu, which arrived late on the scene without an established name and financial backing, secured funds by issuing public stock and borrowing from outside banks. The enormous capital requirements due to the extensive facilities, expensive research and development, and delayed return on investment, made it difficult for the new zaibatsu to put up sufficient private capital. Noguchi, Mori, and Nakano turned to the Industrial Bank of Japan for funds. In addition, Noguchi and Mori borrowed additional funds from Mitsubishi and Yasuda Bank, respectively. By 1936, Nissō was heavily indebted to the Industrial Bank of Japan. The bank sent its own personnel to inspect Nissō's affairs and subsequently demanded Nakano's resignation in 1940.[48] The willingness of the new zaibatsu to take on greater financial risks made them more vulnerable to the whims of the market and the demands of outside investors.

Nissan was unique in being a publicly owned corporation. Ayukawa established Nissan as a public holding company in 1928, after assuming control of his brother-in-law's company, Kuhara zaibatsu, and merging it with his own firm, Tobata Casting. Nissan raised funds for its various subsidiaries through public stock offerings. The company made huge profits from the public stock offering of its mining subsidiary, Nippon Mining, at the time of Japan's reimposition of the gold embargo in the early 1930s. Nippon Mining held a dominant stake in Japan's gold and copper mining; when gold prices rose, its mining stocks increased tenfold.[49] Through a series of timely public offerings of shares of his subsidiaries, Ayukawa was able to raise substantial funds to acquire additional companies. As a result of their firm's large capital requirements and related risks, the new zaibatsu looked to the state as a potential ally in their drive to build up their technological concerns.

---

47. See Masaki Hisashi's discussion of the term in his "The Financial Characteristics of the Zaibatsu in Japan: The Old Zaibatsu and Their Closed Finance," in Keiichirō Nakagawa, ed., *The International Conference on Business History,* vol. 3, *Marketing and Finance in the Course of Industrialization* (Tokyo: University of Tokyo Press, 1978).

48. Arisawa, *Shōwa keizaishi,* 101–102.

49. Ayukawa Yoshisuke, "Nissan kontsuerunu no seiritsu," in Andō Yoshio, ed., *Shōwashi e no shōgen,* vol. 2 (Tokyo: Hara shobō, 1993), 116–117.

These industrialists represented a new type of technocrat who combined the best of both the engineer-scientist and business entrepreneur. Ayukawa, Noguchi, and Ōkōchi graduated from the engineering faculty of Tokyo Imperial University. Ōkōchi later became a professor of engineering at the university. Nakano worked as a research assistant in the science and engineering department at Kyoto Imperial University. Although the new zaibatsu were technology driven, they distinguished themselves not by their technology, but by their ability to find practical, commercial applications for it. One journalist described Ayukawa as a new type of industrialist who "creates, buys, and sells things from the standpoint of an engineer. He does not need to turn commercial practices to his advantage, nor follow the beaten track of industrialists."[50] Moreover, the journalist pointed out, "one cannot attribute this solely to his technical background. There are countless numbers of industrialists with technical backgrounds like Ayukawa. But most of them, before they even realize it, compromise their identity as engineers."[51] The new zaibatsu industrialists were distinguished by their ability to combine a technological vision with the will and organizational and financial resources to implement that vision.

Their record of succeeding where others had failed was attributed to their ability to combine an eye for technological innovation with superior management strategies. The new zaibatsu were keen to apply the latest American management principles and production techniques to enhance their competitiveness. Ayukawa found in Fordist mass production techniques a way to cheaply and efficiently produce his Dat cars. He was also influenced by the concept of stratified holding companies advanced by the utilities magnate Samuel Insull.[52] Ōkōchi was influenced by Taylorist ideas of scientific management in advocating a high level of specialization in the production process.

Both the old and new zaibatsu justified their business activities in patriotic terms and claimed to pursue a higher calling that went beyond the pursuit of profits. From the early years of Meiji, zaibatsu leaders appealed to the traditional Japanese values of diligence, self-sacrifice, harmony, and Confucian paternalism.[53] Toward the political elite they presented themselves as Restoration patriots (shishi), who were implementing the ideals of the Meiji Restoration in business. Toward their employees they adopted a paternalistic stance and exhorted them to put aside private greed and serve the nation. The new zaibatsu, in contrast,

---

50. Wada Hidekichi, "Kaiketsu Ayukawa Yoshisuke," Nihon hyōron (July 1936), 421.
51. Ibid.
52. Nakagawa, "Business Strategy and Industrial Structure in Pre-World War II Japan," 30.
53. On zaibatsu nationalism, see Byron K. Marshall, Capitalism and Nationalism in Prewar Japan: The Ideology of the Business Elite, 1868–1841 (Stanford: Stanford University Press, 1967), esp. 30–50.

promoted a different type of "management nationalism" that embraced the military's strategic goals to build a national defense state and achieve autarky.[54] They portrayed their businesses as oriented toward national over private interests and driven by the logic of technology, rathar than profits.

The new zaibatsu carefully cultivated this techno-nationalist image to the public in their choice of company names. They generally stayed away from family names, as in the case of Mitsui, Yasuda, Nomura, and Suzuki. They selected names that described their technological focus and often with "Japan" or the era name "Shōwa" attached. For instance, Ayukawa named his holding company Nissan (Japan Industries) and called its subsidiaries Japan Mining and Hitachi (Rising Sun) Electric Power. Noguchi adopted the name Japan Nitrogeneous Fertilizer (Nitchitsu), while Nakano named his firm Japan Soda (Nissō). The Mori Concern, which adopted the family name, called its flagship company Shōwa Denkō, which was formed out of the merger of Shōwa Fertilizer and Japan Electric.[55]

The new zaibatsu struck new chords in their critique of liberal capitalism. Ōkōchi promoted his version of "science-based industry" (kagakushugi kōgyō) or "knowledge-based industry" (chinōshugi kōgyō) in contrast to "capitalist industry" (shihonshugi kōgyō).[56] He distinguished Riken from traditional concerns such as the German wartime empire of Hugo Stinnes, who pursued profits through questionable financing and reckless expansion.[57] Ōkōchi argued that at Riken, science and technology, not profits, were the guiding principles. As a result Riken bore a much heavier burden than other companies due to the considerable difficulties and financial risks involved in commercializing its inventions.[58] At the same time, Ōkōchi acknowledged that profits provide the means to pursue its mission of funding scientific research and commercializing inventions. Ōkōchi did not criticize profits but rather the inefficient and rapacious methods by which profits were obtained. In a not-so-subtle criticism of both the old-line and Taishō zaibatsu, he rejected the typical capitalist strategy of maximizing profits by exploiting low-wage labor. The results of such a strategy were, contrary to popular thinking, inefficiencies, waste, higher production costs, and low-quality goods. Ōkōchi argued that what was needed was sophisticated machinery that reduced

---

54. Udagawa, Shinkō zaibatsu, 257.

55. Ibid.; also see Uno Kazushiro, "Ayukawa, Mori, Nakano, Noguchi," Jitsugyō no Nihon (January 1938), 166–167.

56. Ōkōchi's organization also published the journal Kagakushugi kōgyō, through which he and other technocrats promoted their ideas. For an analysis of Ōkōchi's thought, see Ōno Eiji, "Shinkō zaibatsu no shisō," in Chō Yukio and Sumiya Kazuhiko, eds., Kindai Nihon keizai shisō, vol. 2 (Tokyo: Yūhikaku, 1971), 109–128.

57. Ōkōchi Masatoshi, "Riken kontsuerunu no shimei," Kagakushugi kōgyō (June–July 1937), 60.

58. Ibid., 61.

costs and waste, produced high-quality goods, and sustained higher wages. By placing industry on a scientific basis and shifting its priority from capitalism to scientism, Japan would reduce its costs; increase its wages; and create original, superior products—a vision that he believed Henry Ford had realized in his mass production of cars.[59] Ultimately such a strategy would strengthen Japan's export capabilities.

Ayukawa also promoted a more public-spirited capitalism through his idea of the public holding company (*kōkai mochikabu gaisha*). Ayukawa claimed that, unlike the traditional zaibatsu leaders, he did not seek to amass personal wealth. He criticized the "family" (*ie*) tradition of zaibatsu such as Mitsui and Mitsubishi, in which businesses were passed down to sons and adopted sons. Ayukawa's philosophy was that "you live your own life, accomplish what you can, and 'that's it.'"[60] Like other military leaders, he harbored a distrust of financial and economic theory and denounced classical economics for being too abstract.[61] One journalist noted that "the essence of economics is its lack of conclusions; from the standpoint of the average industrialist, who manipulates a world without conclusions to his own advantage, Ayukawa is certainly an unusual type."[62] He was described as one who prefers to deal with "things" instead of abstract ideas.[63] Ayukawa saw himself as a servant of public stockholders who uses their funds to create profits in the form of dividends, capital gains, and "concrete" tangible products.[64]

In contrast to the traditional zaibatsu's uncompromising stance toward organized labor, the new zaibatsu sought to appeal to labor by promoting a new productivist vision of industry in which technology, not profits, is the driving force. They argued that Japan should create high-quality, high-value-added goods through superior technology and efficient, low-cost production, instead of shoddy, low-margin goods through the exploitation of cheap labor. Like the advocates of scientific management and industrial rationalization, they appealed to technology to recast industrial relations from an entrenched, conflict-ridden relationship, in which capitalists profit at the expense of workers, to a dynamic, cooperative relationship, in which workers and capitalists strive together to increase the overall profits of the firm *and* sustain higher wages.[65] Neither left nor

59. Ibid., 69–70.
60. Ayukawa, "Nissan kontsuerunu no seiritsu," 113.
61. Wada, "Kaiketsu Ayukawa Yoshisuke," 422.
62. Ibid.
63. Ayukawa, "Nissan kontsuerunu no seiritsu," 125.
64. Ibid.
65. William M. Tsutsui, *Manufacturing Ideology: Scientific Management in Twentieth-Century Japan* (Princeton: Princeton University Press, 1998).

right, their vision of modern, technology-based industry transcended both the traditional class orientation of the left and Japanist appeals to time-honored traditions of the right.

Reform Bureaucrats and the Managerial State

The bureaucracy has been commonly viewed as an undifferentiated and monolithic force in Japanese politics. Its ethos and mode of operation have been understood in the Weberian terms of a status-bound, rule-based culture focused on enforcement. Japan's bureaucracy, as other modern bureaucracies, was designed to safeguard the state's authority and legitimacy through the uniform interpretation and application of laws within a fixed legal structure; its mandate was administration, not innovation.[66] Under this system, legal training was favored over technical training. The bureaucracy took its final form in the early 1900s, at a time when it faced challenges from outside political groups. As the influence of the Meiji oligarchs waned, the bureaucracy sought to acquire a monopoly over legal expertise and "publicness" by establishing a routinized career structure that included graduation from the law faculty of Tokyo Imperial University, the passing of a rigorous civil service exam, and promotion based on specialization and seniority.[67] During the Meiji and Taishō periods, the bureaucracy aimed to foster the conditions for public order, social harmony, and economic growth. Japan's higher civil servants shared a strong elite consciousness as "officials of the emperor" and sense of duty as "shepherds of the nation" to promote public over private interests, protect the weak from the strong, and mediate between class and sectoral interests.[68] After the bureaucracy helped establish the first factories and nurture its export industries under the policy of "industrial promotion," it gradually privatized industries and adopted a more laissez-faire approach toward business. At the same time, it devoted considerable energies to delimit political participation and suppress the left wing.

66. Bernard S. Silberman, "The Bureaucratic Role in Japan, 1900–1934: The Bureaucrat as Politician," *Japan in Crisis* (Princeton: Princeton University Press, 1974), 207; Gouldner, *Dialectic of Ideology and Technology,* 250–274.

67. Bernard S. Silberman, "The Bureaucratic State in Japan: The Problem of Authority and Legitimacy," in Tetsuo Najita and J. Victor Koschmann, eds., *Conflict in Modern Japanese History* (Princeton: Princeton University Press, 1982).

68. On Japanese bureaucrats and social policy, see for example Kenneth Pyle, "Advantages of Followership: German Economics and Japanese Bureaucrats," *Journal of Japanese Studies* (autumn 1974), 127–64; Sheldon Garon, *The State and Labor in Modern Japan* (Princeton: Princeton University Press, 1987).

From the late 1920s, however, a group of technically minded bureaucrats began to view the traditional conception of bureaucracy as outdated and incapable of meeting the challenges of the modern era. These so-called "reform bureaucrats" sought to inject a new dynamism and purpose into the bureaucracy and profoundly alter its culture and spirit. They advanced a new technocratic vision of a "managerial state" (*kanri kokka* or *keieisareta kokka*) in place of the "night watchman state" (*yoban kokka*) of the past.[69]

The Japanese state faced new administrative challenges in response to the technicization of war, industry, and society. Particularly after the invasion of Manchuria, the state assumed the central role in total war mobilization, resource development, and managing an expanding empire. Reform bureaucrats criticized the bureaucracy's traditional supervisory, regulatory role as a night-watchman state that focused primarily on rule enforcement and efficient administration. They argued that the increasingly complex and technical nature of government required a more activist, interventionist, and technically sophisticated state. They sought to reinterpret the bureaucracy's mandate from the uniform and impartial application of law to the efficient and effective management of Japan's emerging advanced national defense state (*kōdō kokubō kokka*). The new mandate required active cooperation and collaboration among the bureaucracy, business, and the military across the jurisdictional boundaries dividing public and private and military and civilian affairs.

Reform bureaucrats called on their colleagues to change their attitude of passive conformity and focus on narrow sectional interests to a proactive stance in which bureaucrats assume a sense of responsibility and take decisive action with a feeling of duty to the state. They called for the reeducation of officials to provide them with specialized training and a sense of political accountability, promotion based on the criteria of efficiency and results, and preferential treatment toward those with technical expertise and experience. Particularly with regard to economic affairs, reform bureaucrats advocated the replacement of traditional "legislative bureaucrats" with a new type of bureaucrat described as "administrative technologists," "creative bureaucrats," or "creative engineers" who approach their administrative duties in terms of active management (*keiriteki kinō*) rather than of passive supervision (*kantokuteki kinō*) as in the past.[70]

These bureaucrats believed that technology was introducing a new managerial logic into the bureaucracy. The administrative science expert Rōyama Masamichi argued that in Japan, technological advance was bringing about the

---

69. Okumura Kiwao, *Nihon seiji no kakushin* (Tokyo: Ikuseisha, 1938), 133.

70. Kashiwara Heitarō, Minobe Yōji, Sakomizu Hisatsune, and Mōri Hideoto, "Zadankai: kakushin kanryō—shintaisei o kataru," *Jitsugyō no Nihon* (January 1941), 54.

technicization of bureaucratic administration by which the traditional administrative functions were evolving toward the management functions of private industry.[71] According to Rōyama, during the nineteenth-century, new mechanical devices, such as the adding machine and telegraph, and new sciences, such as civil engineering, health administration, and accounting, had increased the geographical and functional scope and efficiency of the bureaucracy. After World War I, this "material technology" gradually began to induce a qualitative change in administrative ability by bringing about the technicization of the administrative function itself. Rōyama viewed bureaucratic administration as a form of technology, which he defined as "the application of knowledge to achieve certain goals."[72] He believed that administration was acquiring functional, goal-oriented characteristics by utilizing various methods or management elements to attain state goals. Although the bureaucracy did not determine the goals itself—these are determined in the political arena—it exercised decision-making power over the means to achieve them. Rōyama referred to this "system of practical, task-oriented decision-making" as "management" (*kanri*).

> Moreover, although management is not a technology itself, it is very much a technological matter. That is, the implementation of specific government goals through the employment of immense numbers of people, the expenditure of hundreds of millions [of yen] worth of funds and goods, and the planning and operation of complex organizations, which [together] represent one administrative task itself, is a problem that requires independent, technological deliberation. It is clear that the assembly of these various elements, such as personnel, funds, materials, and organizations, and their use for established objectives forms one system of action-oriented decision-making. State administration today is in a state of rapid reorganization, training, and retooling of management elements which are already rapidly developing phenomena of business management and administration in private firms. Since the European war and particularly in today's era of reform, the problem of administration is none other than the problem of management.[73]

The increased need for technocratic expertise in military and economic affairs raised the status and influence of technically oriented bureaucrats and their affiliated ministries and planning bureaus within the government. Modern technology and the new developments in science-technology elevated the

---

71. Rōyama Masamichi, "Gyōsei to gijutsu," *Kagakushugi kōgyō* (May 1938), 13–23.
72. Ibid., 18.
73. Ibid.

image of technology from an artisan's trade into something more respectable, rational, and socially relevant.[74] Moreover, this new techno-scientific complex and its technical personnel represented a new technical subsystem and planning network within the bureaucracy.[75] In Japan, these supraministerial planning bureaus and agencies enabled technocrats to gradually increase their control over strategic, technology-related areas such as materials production and planning, energy resources, foreign trade, communications, and propaganda both at home and in Manchuria. Furthermore, they were empowered by a series of unprecedented mobilization and control laws backed by the force of the state.

Reform bureaucrats operated at the heart of the state's expanding industrial and technical planning network. They were predominantly young career bureaucrats from the technical ministries and cabinet planning agencies whose technocratic orientation and expertise made them compatible partners to military planners. Some of the most influential technocrats worked at the Ministry of Commerce and Industry, which played a leading role in the military's war mobilization program.[76] The Commerce Ministry's mandate to develop, improve, and promote industry encouraged an innovative approach toward economic policy that the military required. During the war, this dynamic approach became especially important when foreign trade declined and the economy shifted to a semiautarkic, materials-based economy. In contrast, the Finance Ministry, given its role of securing and distributing public funds, tended to value economic stability over innovation. Its fiscally conservative stance often conflicted with the military's demands for increased budgets. Kishi Nobusuke rose to the ranks of vice minister and minister of commerce, and later vice minister of munitions during the early 1940s. His mentor, Yoshino Shinji, served as vice minister and minister of commerce between 1931 and 1938. Kishi's protégé Shiina became vice minister of munitions in the early 1940s. At the Cabinet Planning Board, the top civilian planners included Hoshino Naoki, Minobe Yōji, Mōri Hideoto, Okumura Kiwao, and Sakomizu Hisatsune. Unlike Kishi, Shiina, and Yoshino these men did not rise to the top posts within their own ministries; most achieved prominence through their activities at supraministerial agencies and served as important conduits between the Cabinet Planning Board and their home base. Closely affiliated with this group were young left-wing planners such as Wada Hiroo, Kashiwara Heitarō, Katsumata Seiichi, and Inaba Hidezō.[77]

---

74. Gouldner, *Dialectic of Ideology and Technology*, 251–254.

75. Ibid.

76. On the history of the Ministry of Commerce and Industry, see Johnson, *MITI and the Japanese Miracle*.

77. For a study of these left-wing technocrats, see Laura Hein, *Reasonable Words, Powerful Men* (Berkeley: University of California Press, 2004).

The reform bureaucrats formed a cohesive group based on their university ties and shared professional training at the technical ministries, cabinet research agencies, and planning departments of Japan and Manchukuo. In terms of their formal education, most studied law at Tokyo Imperial University.[78] Already at the university, a certain technocratic orientation informed their rather atypical choice of bureaucratic career paths. Many opted for the technology-related "second tier" ministries instead of the prestigious Home and Finance ministries, because they believed that the technical ministries would play a pivotal role in the future. Kishi, a top student of German law, chose the Commerce and Agriculture Ministry over the Home Ministry. He later claimed that his decision was based on the belief that:

> As for the future of Japan, in order for resource-poor Japan to maintain itself, it must establish itself through trade. In order to establish itself through trade, industrial technology needs to be developed, and through technological superiority, industry must be developed. The Ministry of Commerce and Agriculture was in charge of that administration.[79]

Okumura justified his decision to join the less prestigious Ministry of Communications in terms of his interest in the state's social role, particularly the administration of public utility enterprises.[80] University ties provided these men with lifelong personal networks and an elite esprit de corps that was further strengthened by their new identity as leaders at the forefront of technology.

More than their alumni networks, their shared practical training in research, planning, and implementation throughout the 1930s and early 1940s fostered close working relationships and a common technocratic approach. The financial crisis, domestic right-wing terror, and the military's expanding presence in north China prompted the emergence of military-bureaucratic cabinets and the need for active and continuous collaboration across ministerial lines. From 1932, reform bureaucrats were sent to Manchuria to draft and deliberate policies under the "internal guidance" of army technocrats. The experience of joint planning among civilian and military technocrats in Japan and Manchuria gave rise to a new conception of state-society relations and calls for a more activist state to prepare for total war.

---

78. Two exceptions were Mōri Hideoto, who specialized in political science at the law faculty of Tokyo Imperial University, and Inaba Hidezō, who studied philosophy at Kyoto Imperial University.

79. Kishi Nobusuke, "Shisei de ugokanu mono wa nai: Watakushi no jinsei o kettei zuketa Shōin sensei no kotoba." Kishi's speech at the Bunyūkai on May 9, 1980, 6. Kishi graduated at the top of his class in the law faculty at Tokyo Imperial University and was a favored student of Uesugi Shinkichi, who invited him to stay on at the university and pursue an academic career.

80. Okumura Kiwao, *Teishin ronsō* (Tokyo: Kōtsū kenkyūsha, 1935), 1.

Like their counterparts in the military and industry, the reform bureaucrats were above all pragmatists, not ideologues. Their vision did not derive from one particular intellectual source. In their writings and drafts they drew on a diverse body of work including Marxist theory, Soviet planning models, New Deal writings, Nazi political-economic theory, and Japanese right-wing tracts. The group's eclectic approach reflected the diverse intellectual backgrounds, personal styles, and functional roles of its members. The group's acknowledged ideologues, Mōri Hideoto and Okumura Kiwao, formulated the basic principles and conceptual framework for planning. Both men were intellectual-bureaucrats who were conversant in Western political-economic theory and published their writings in professional and mass journals. The more centrist, business-minded members, such as Kishi, Shiina, Minobe, and Sakomizu, acted as public spokesmen for the group and took charge of implementation. They translated the technical language of planning to the public and mediated between the military, bureaucracy, and business. The left wing included economic researchers such as Wada Hiroo, Inaba Hidezō, and Katsumata Seiichi.

Their early intellectual training and political background informed their diverse approaches toward planning. At the university, most reform bureaucrats studied either German or English law. Those who studied German law, such as Kishi, Shiina, and Okumura, tended toward a more statist approach. Kishi was a protégé of the right-wing legal scholar Uesugi Shinkichi. Both Kishi and Okumura drew directly on German models of economic and political mobilization based on their firsthand observations of German political and industrial organizations in the 1930s. In contrast, students of English law, such as Minobe, Sakomizu, and Wada incorporated progressive, socialistic approaches into their state-centered views. In their youth they flirted with left-wing politics and developed their theoretical skills by studying Marxism.[81] They cultivated their interest in academic Marxism through participation in left-wing student organizations such as Yoshino Sakuzō's Shinjinkai, social welfare projects such as the Yanagishima Settlement, and sports clubs.[82] Although they abandoned their social activism on entering the ministry, they retained a penchant for Marxist theoretical reasoning in their planning: these bureaucrats tended to

---

81. Furukawa Takahisa, "Kakushin kanryō no shisō to kōdō" *Shigaku zasshi* (April 1990), 1–38; Furukawa Takahisa, *Shōwa senchūki no sōgō kokokusaku kikan* (Tokyo: Yoshikawa kōbunkan, 1992). On the influence of Marxism on Japanese bureaucrats and intellectuals, see Hein, *Reasonable Men, Powerful Words;* Bai Gao, "Arisawa Hiromi and His Theory of the Managed Economy," *Journal of Japanese Studies* 20, no. 1 (winter 1994), 115–153; and Gao, *Economic Ideology and Japanese Industrial Policy* (New York: Cambridge University Press, 1997).

82. Nihon hyōron shinsa, ed., *Yōyōtaru: Minobe Yōji tsuitōroku* (Tokyo: Nihon hyōron shinsha, 1954), 328–332.

view individual events within broader historical processes and adopted a holistic view of political-economy as an organic, integrated system with its own iron laws and internal logic. Colleagues of Minobe attributed his remarkable grasp of the problems of the controlled economy and skill in planning and implementation to his solid foundation in Marxist theory.[83] Their intellectual approach was similar to that of the new generation of military planners who had also studied Marxism such as Suzuki Teiichi, Akinaga Tsukizō, and Ikeda Sumihisa, and left-wing economists such as Arisawa Hiromi, Ōuchi Hyōe, and Ryū Shintarō. Although they drew on Marxist conceptual tools and embraced the socialist faith in state planning, they rejected the theory of class conflict. By the late 1930s reform bureaucrats on the left, right, and center argued that the motor force of history was neither class struggle nor individualism, but technology and the *Volk*. They believed that the new era was moving beyond the materialism and class orientation of liberalism and Marxism and embracing a third way that they associated with European fascism.

The National Economy

By the time of the Great Depression, neither the left nor labor posed a serious challenge to the state. The Communist Party had been driven underground since 1927, and labor issues took a back seat to the more pressing demands of the rural crisis, economic recovery, and war mobilization. The moderate Sōdōmei-led trade union suffered a rapid decline after making important legislative gains throughout the 1920s. In its place emerged increasingly vocal right-wing ultra-nationalist, Japanist, and pan-Asianist groups. In the new climate of political reaction and war fever, labor joined hands with the state. The recently formed Social Masses Party promoted corporatist notions of an organic social order that denied the conflict of labor and capital.

With the rise of the managerial state, political struggles in the workplace between management and labor were overshadowed by new battles at the industry and national level between big business and the state. The main challenge faced by reform bureaucrats was to obtain the cooperation of big business. With the onslaught of the world depression, large firms in industries such as shipbuilding, electrical machinery, and chemicals began to engage in destructive price cutting, even selling below cost, in order to maintain international market share. The Commerce Ministry proposed the formation of industrial cartels modeled on those created for small- and medium-sized export firms in 1925.[84] Under

---

83. Ibid., 325–326, 342–343.
84. Johnson, *MITI and the Japanese Miracle,* 98–99.

the Exporters Association Law, the state established export unions for various product lines and an inspection system to control the quality, price, and quantity of export goods. Under the Major Export Industries Association Law, the state succeeded in inducing small- and medium-sized firms to coordinate production and pricing by offering government subsidies and facilities for cartel members.

In contrast to the positive response of small- and medium-sized firms to state efforts, however, the ministry faced considerable resistance from big business. Executives complained that the ministry's rationalization policies came too late. Had such measures been promoted in the initial phases of depression, they argued, their firms would have welcomed assistance. Since they had already formulated their own response to the crisis, they were not willing to adopt a new course. Moreover, they pointed out, given the government's tight money policy, it clearly did not have adequate funds to support the type of overall restructuring needed.[85]

As reform bureaucrats soon realized, unlike the small- and medium-sized firms, the heavy industrial concerns were fewer in number and wielded more power within their respective industries. The considerable expertise and hubris of their leaders and their extensive personal ties within the business and financial community made them less willing to accept guidance from bureaucrats. At the same time, Commerce Ministry officials feared that if they allowed the present events to run their course, these firms would end up hurting not only themselves, but also their industries, workers, raw materials suppliers and distributors, financial institutions, and investors. The root of the problem, as Commerce Minister Yoshino Shinji saw it, lay in the system of liberal capitalism and its short-sighted emphasis on profits and free competition.[86]

Yoshino assigned his protégé Kishi the task of formulating an appropriate response. Reflecting on his firsthand investigations of Germany's industrial rationalization movement in 1927 and 1930, Kishi advanced two basic principles. The first was that cost efficiency, or the minimization of input per unit of output, should be the preferred method of achieving profits. He criticized the American approach of increasing profits by engaging in wasteful and inefficient practices such as fixing prices, undercutting competitor's prices, and artificially limiting supply.[87] He believed that firms could become more profitable if they systematically applied the new management technologies such as scientific management,

85. Yoshino Shinji, "Sangyō gōrika," in Andō Yoshio, ed., Shōwashi e no shōgen, vol. 1 (Tokyo: Hara shobō, 1993), 172–174.

86. Yoshino Shinji, Nihon kōgyō seisaku (Tokyo: Nihon hyōronsha, 1935), 320.

87. Kishi, "Ōshū ni okeru sangyō gōrika no jissai ni tsuite," Sangyō gōrika (January 1932), 31–32.

mass production, and techniques to streamline the production process and distribution networks.

The second principle was the importance of national cooperative effort. Kishi praised the German notion of "national cooperative effort" or "community of work" (*Gemeinarbeit des Volks, Gemeinwirtschaft, Gemeinschaftsarbeit der ganzen Wirtschaft*), which he believed reflected the "true spirit" of German rationalization.[88] For Kishi, the concept represented a new way of thinking about the economy. In contrast to the liberal mentality of profit for profit's sake, it emphasized order, efficiency, and cooperation. This corporatist vision promoted the idea of an organic, hierarchical, functionalist society comprised of occupational estates, rather than a class-based society made up of winners and losers. It was employed in the early 1900s by German heavy industry to counter the class-based challenges of labor.[89] By the late 1920s it was taken up by semiofficial organizations such as Germany's rationalization board, the Reichskuratorium für Wirtschaftlichkeit (RKW), to promote cooperation among firms.[90]

Kishi was impressed by the RKW's promotion of "experience exchange" (*Erfahrungsaustausch*) in which firms joined industry-based groups and shared information about their production, management, technology, and sales strategies.[91] In contrast to the laissez-faire system of free competition in which firms closely guarded information about their operations and managed their businesses in isolation, the new system encouraged firms to disclose their management secrets with the aim to improve their own techniques by learning from their competitors.[92] Kishi also praised the German strategy of comparing business results (*Betriebsvergleich, jigyō hikaku*) by which firms disclose proprietary data on output, expenses, and sales to a third party. This intermediary party then ranked each firm's performance based on the average figures of the group and dispatched consultants to improve each firm's business strategy. He proposed

---

88. Kishi, Yatsugi, and Itō, "Kankai seikai rokujūnen: Dai-ikkai Manshū jidai," 293; Kishi, "Ōshū ni okeru sangyō gōrika no jissai ni tsuite," 30.

89. See Dennis Sweeney, "Corporatist Discourse and Heavy Industry in Wilhelmine Germany: Factory Culture and Employer Politics in the Saar," *Comparative Studies in Society and History* 43 (October 2001), 701–734.

90. On the RKW, see Robert E. Brady, *The Rationalization Movement in German Industry: A Study in the Evolution of Economic Planning* (Berkeley: University of California Press, 1933); Mary Nolan, *Visions of Modernity* (Oxford: Oxford University Press, 1994); J. Ronald Shearer, "The Reichskuratorium für Wirtschaftlichkeit: Fordism and Organized Capitalism in Germany, 1918–1945," *Business History Review* 71, no. 4 (winter 1997), 569–602; Dr. Georg Freitag, "Das Haus Siemens und das RKW: Ein Beitrag zur Gründung und Entwicklung der deutschen Rationalisierungsbewegung," *Rationalisierung* 14 (1963).

91. *Reichskuratorium für Wirtschaftlichkeit, RKW Nachrichten* (September 1931).

92. Kishi Nobusuke, "Sangyō gōrika undō ni arawaretaru keiken kōkan," *Kōgyō keizai kenkyū* (July 1932), 103.

that such measures be adopted by Japan's Temporary Industrial Rationality Bureau, which was modeled on the RKW.

Both the principles of "national community of work" and cost efficiency formed the cornerstone of the concept of the "national economy" (*Volkswirtschaft, kokumin keizai*). Originally derived from Friedrich List's mercantilist concept of economic nationalism and competition among states, the vision of national economy acquired a more pronounced technocratic meaning by the 1930s.[93] The RKW described Germany's national economy as being constituted on the technical (intrafirm), commercial (intraindustry), national, and international levels and based on the four principles of systematic order, efficiency, statistical comparison, and societal well-being.[94] Similarly, the reform bureaucrats defined Japan's national economy in terms of the application of science and maximum cooperative effort—or systematic, technical, communal work—for the purpose of achieving a more rational organization of the economy and increasing society's well-being. National economy promoted outward competition among states and inward cooperation among domestic firms. But its main focus was on achieving optimal balance and coordination between and among the various, often conflicting, levels and means of rationalization. National economy signified a highly integrated, coordinated, functionally interdependent economy, a type of "corporatist economy" (*korporative Wirtschaft*) or "confederational economy" (*Verbundwirtschaft*) that required long-range planning by firms, intermediary organizations, and the state.[95]

National economy served as the guiding concept under which bureaucrats formulated their first control policy toward big business. Under the Important Industries Control Law of April 1, 1931, if more than two-thirds of the firms within an industry joined a cartel and that cartel was approved by the state, the state could enforce the participation of outside firms in the cartel. The state would authorize a cartel if it determined that its activities promoted the "healthy development of the national economy" and "protected public interests."[96] In cases where the state found that the activities of the cartel violated public interests, it had the right to modify or abolish the cartel, as well as impose fines and penalties on its violators. In terms of the purpose of the law—the formation of cartels

---

93. Kishi, "Ōshū ni okeru sangyō gōrika no jissai ni tsuite," 34.

94. *Reichskuratorium für Wirtschaftlichkeit, RKW Nachrichten* (July 1931), 109. See also Friedrich von Gottl-Ottlilienfeld's *Vom Sinn der Rationaliesierung* (Jena: Gustav Fischer, 1929), which provided one theoretical basis for the RKW's activities.

95. Robert Brady, *The Rationalization Movement in German Industry*, 7, 362, 395; *Reichskuratorium für Wirtschaftlichkeit, RKW Nachrichten* (August 1931), 127.

96. Kishi Nobusuke, "Jūyō sangyō tōsei hō kaisetsu," *Kōgyō keizai kenkyū* (April 1932), 51–76.

among the major firms—it represented a continuation of the ministry's cartel policy—but now enforced by the state.

Yoshino emphasized after the war that there was no intent to establish a controlled or planned economy at the time.[97] Nevertheless, important precedents were established. From the legal point of view, the nature of industrial cartels changed from private entities based on private contract, to public entities based on state law.[98] Whereas cartel policy before the depression was promoted within the liberal capitalist framework of laissez-faire and the protection of private property, now cartels were to be administered from the standpoint of the national economy. Kishi pointed out that the new law was the first to use the term *national economy* in legal language, and that it was conceived specifically with a critical view toward liberalism.[99] The Important Industries Control Law was only the first of a series of state control measures. More than its modest goals, its underlying principle of national economy introduced an important conceptual basis upon which to justify state intervention into the affairs of private enterprise and alter the relationship between the public and private sector.

The technocratic visions and strategies of total war officers, new zaibatsu, and reform bureaucrats revealed a new, modern mindset. Their critique of liberalism was not a rejection of modernity and a reactionary turn to the past but rather a quintessentially modernist act. The rise of technology brought about the technicization of not only war, industry, and administration, but also of Japanese professionals themselves. As modernists, they became both the subjects as well as objects of technology; they embraced technology and in turn were transformed by it.

The wartime collaboration among technocrats was, to a certain extent, a marriage of convenience—a unique opportunity to influence government policy. But these groups also shared a broad basis for cooperation in terms of a similar worldview and common language of technocratic planning. First, despite their different professional concerns and interests, these groups tended to focus on the interrelatedness of various processes. In order to prepare for total war, military officers aimed at comprehensive mobilization of the nation's material and spiritual resources. New zaibatsu leaders sought to integrate various aspects of production in technologically related areas. Within the framework of the managerial state, reform bureaucrats attempted to coordinate the development and

---

97. Yoshino, "Sangyō gōrika," 176.
98. Homma, "Senji keizai hō no kenkyū," 6–15.
99. Kishi, "Sangyō gōrika yori tōsei keizai e," *Sangyō gōrika* (April 1934) 33; Kishi, "Jūyō sangyō tōsei hō kaisetsu," 54.

administration of important industries and social services via state guidance and control. Whereas in the past firms had viewed each other as competitors, now they should join together and compete as one nation. Bureaucrats should overcome sectionalism and red tape and strive toward efficient and effective administration. Second, the various strategies of these groups reflected a more general shift in focus from specific processes and actors to metaprocesses and broader units: not just troops, but nations would fight future wars; not individual businesses, but a number of technologically linked businesses were combined under one firm; and not the individual firm or ministry, but the economy and bureaucracy as a whole became the defining unit. Third, they formed their critical stance toward Japan's liberal capitalist system from the standpoint of technology and state nationalism. Military leaders viewed the profit orientation and laissez-faire approach as an obstacle to national mobilization. New zaibatsu leaders set as their overall goal the advancement and application of public technology instead of the preservation of private wealth. Reform bureaucrats viewed sectionalism and narrow specialization within the bureaucracy as short-sighted and against the national interests. Moreover, these leaders tended to view themselves as transcendent technocrats and nationalists who stood above private interests and represented the interests of the whole. Both total war officers and reform bureaucrats believed that they possessed the technical expertise and insight to command greater cooperation within and among various industrial sectors, ministerial departments, and private groups, while new zaibatsu leaders believed themselves to be the servants of the public interest and of Japanese science-technology in the modern era.

Finally, in their visions of technocratic modernity, these leaders sought a third way beyond liberal capitalism and socialism that drew on both left- and right-wing critiques of capitalism. Japanese technocrats viewed the Marxist class-based critique of the left and the restorationism and Japanism of the right as outdated and ineffective because both failed to account for the rise of technology. But they selectively embraced aspects of the left-wing populist, socialist vision and faith in state planning and the right-wing appeal to Japanese tradition and state and ethnic nationalism and incorporated them into their own technocratic agendas. By sublating the anticapitalist agendas of the left and right, technocrats helped reshape the political landscape of interwar Japan. By the mid- to late 1930s the main battle lines in policymaking were drawn no longer along the horizontal axis of left, right, and center or progress versus reaction, but along the vertical axis of liberal capitalist status quo (*genjō iji*) versus reform (*kakushin*).[100]

---

100. Itō Takashi, "The Role of Right-Wing Organizations in Japan," in Dorothy Borg and Okamoto Shumpei, eds., *Pearl Harbor as History: Japanese-American Relations, 1931–1941* (New York: Columbia University Press, 1975), 487–491.

2

MILITARY FASCISM AND MANCHUKUO, 1930–36

The Great Depression marked the ascendance of the Japanese military and the right wing as a political force in the 1930s. For these groups, the financial crisis and collapse of world trade signified the implosion of the liberal capitalist system. Prime Minister Hamaguchi Osachi's retrenchment policy and ill-fated decision to lift the gold embargo in 1929 led to one of the worst economic crises in Japanese history.[1] The painful effects of the Minseitō government's tight money policies, its inability to pull Japan out of the depression, and the series of corruption scandals involving business provided ammunition for the right wing's attack on Japan's capitalist system and the ruling establishment. As part of his retrenchment policy, Hamaguchi also pushed through the highly unpopular London Naval Treaty in April 1930. The treaty sought to check the naval arms race by limiting Japan's naval forces to a percentage of that of the United States and Britain. In November of that year, Hamaguchi was fatally stabbed by a member of a right-wing patriotic society. This act was followed by a series of assassinations and coup d'état plots carried out by a group of junior-ranking, radical army and navy officers known as the "young officers." Their reign of terror culminated in the bloody coup d'état attempt on February 26, 1936.

---

1. Mark Metzler, *Lever of Empire: The International Gold Standard and the Crisis of Liberalism in Prewar Japan* (Berkeley: University of California Press, 2006), 217–239; Richard Smethurst, *From Foot Soldier to Finance Minister: Takahashi Korekiyo, Japan's Keynes* (Cambridge, MA: Harvard University Asia Center, 2007), 238–267.

Amidst the domestic crisis, a group of high-ranking army technocrats was instigating a crisis abroad in their resolution of the "Manchuria-Mongolia problem." During the 1920s, Japan's Kwantung Army had become increasingly concerned about the heightened anti-Japanese sentiment in the Manchuria-Mongolia region and its inability to control the Manchurian warlord Zhang Zuolin. In June 1928, Kwantung Army officers murdered Zhang by exploding a bomb on the Japanese controlled Mantetsu railway track as his train car was passing through. Attempts to control Zhang's son, Zhang Xueliang, were no more successful. Ishiwara Kanji, who was serving as a Kwantung Army staff officer, developed a plan for the seizure of Manchuria-Mongolia. On September 18, 1931, Ishiwara and Kwantung Army leaders Itagaki Seishirō and Doihara Kenji staged the bombing of the Mantetsu railway line near Mukden and used this incident as a pretext to invade Manchuria. The Kwantung Army, together with military technocrats in Tokyo, then embarked on an ambitious antiliberal experiment to establish a fascist Manchurian state and planned economy.

The military's two-pronged assault on liberal capitalism set the stage for the rise of fascism in wartime Japan. During the 1930s radical young officers and army technocrats launched a fascist "reform movement" (*kakushin undō*) in Japan and Manchuria. These groups promoted different agendas. The group of radical young officers associated with the army's Imperial Way faction sought to spiritually reform Japan by overthrowing its corrupt capitalist system and eliminating the evil advisors to the throne. The other group of military technocrats affiliated with the Control faction aimed to secure natural resources and develop heavy industry in Manchuria to prepare for a future total war. When these two groups joined together in a temporary, but fateful, alliance in the early 1930s, they promoted fascism in three ways. First, through acts of terror and violence, the young officers brought an end to party government and shifted Japanese politics to the right. Second, military technocrats occupied Manchuria and created a new type of fascist state and economy. Third, these groups paved the way for the rise of a new group of technocratic leaders who embraced and later transformed the military's fascist visions and programs.

Blueprints for Military Fascism

During the 1930s, young officers and military technocrats advanced their own blueprints for military fascism. At the time of the liberal crisis, the young officers had already possessed their own plan based on Kita Ikki's "General Outline of Measures for the Reconstruction of Japan." Kita's outline laid out a radical plan to overthrow the Japanese government and reconstruct it along national socialist lines. Although banned after its appearance in 1919, his political tract circulated widely, not only among young officers, but also elite staff officers and bureaucrats. Kita is considered a symbol of Japanese fascism and his outline as the bible of the young officers. Kita laid out a fascist program for reform that called for an authoritarian state; elimination of class privilege and class conflict; state economic control; expansion abroad; and priority of the ethnic national community over the individual.

Kita offered a penetrating critique of Japan's capitalist state from a national socialist perspective. He argued that the original restorationist vision of close ties between the emperor and the people had been subverted by Meiji oligarchs and capitalists. These leaders erected a capitalist system, founded on private property ownership and class divisions, in which elites manipulated the state to create bastions of privilege and pursue "interest politics." According to Kita, these leaders secured their dominance by promoting the reactionary idea of an unchanging, divine polity headed by an "imperial line unbroken for ages eternal" in which sovereignty lies in the emperor. Kita called for the creation of a "civilian state" (kōmin kokka) in which sovereignty resides in the state, not the emperor, and the emperor and people participate as organs of the state and share national wealth.[2] To realize this vision, he proposed a military coup, temporary suspension of the constitution, and institution of martial law under which the state would eliminate capitalist privilege and promote popular welfare. Reforms included abolishing aristocratic strongholds such as the House of Peers, Privy Council, and Imperial Household; limiting private capital and land ownership, including those of the imperial family; passing a universal suffrage law; improving working conditions; and streamlining and reorganizing industry under centralized state control.

Young officers attempted to implement Kita's vision in a major coup d'état attempt on February 26, 1936. Known as the February 26 incident (ni-ni-roku jiken), it was the last of a series of violent conspiracies to reform the government, which included the October and March incidents in 1931 and the League of Blood incident and May 15th incident in 1932. In this final and most violent assault, fourteen hundred officers occupied key political centers in Tokyo for several days and murdered prominent civilian leaders such as Saitō Makoto and former finance minister Takahashi Korekiyo. Shaken by these uprisings, the government executed eighteen of the leaders, including Kita, whose work was viewed as the ideological inspiration behind the plot. These leaders were

2. On Kita Ikki see George M. Wilson, *Radical Nationalist in Japan: Kita Ikki, 1883–1937* (Cambridge, MA: Harvard University Press, 1969); Christopher W.A. Szpilman, "Kita Ikki and the Politics of Coercion," *Modern Asia Studies* 36, no. 2 (May 2002), 467–490; Kawahara Hiroshi, "Kita Ikki to Shōwa isshin," in Kawahara Hiroshi, *Asia e no shisō* (Tokyo: Kawashima shoten, 1968), 155–163.

condemned by the emperor as traitors to the throne. Moreover, eighteen army generals, many associated with the Imperial Way faction, were forced to retire from active duty and staff officers were transferred from Central Headquarters to regional posts. The February incident brought about the demise of the radical right and the Imperial Way faction and the ascendance of the Control faction within the army.

Military technocrats affiliated with the Control faction set out their own fascist vision for Japan in the so-called "army pamphlet" of 1934 entitled "On the Basic Principles of National Defense and Its Intensification." The slim booklet was one of a series of pamphlets issued by the Army Ministry's press section to explain its national defense policy in Japan and Manchuria and defuse charges that the army was promoting socialist ideas and plotting a violent revolution. The army sought to appeal directly to students and the general reader and solicited their comments by providing a return address. Drafted by officers with strong Marxist backgrounds such as Ikeda Sumihisa and Suzuki Teiichi, the pamphlet set out the Control faction's vision of total war and the necessary national defense measures to prepare for it.

The pamphlet glorified war and made it the basis for state reform. In the opening lines, it described war as "the father of creation and mother of culture." According to the pamphlet, the technical demands of total war brought about the need for a new type of activist, mobilized "national defense state." In such a state the lines between war and peace and public and private were blurred. In the age of total war, societies were in a constant state of mobilization and planning for total war. Hence, the periods between the wars also became a form of war. Moreover, it described national defense, like war, as "a function of the basic vital force of the formation and development of the state." The basic meaning of national defense was that war mobilization played an integral role in the state's development. The pamphlet called for comprehensive state-led mobilization of the material and spiritual resources of the nation and the total reorganization of the Japanese state and society for the purposes of war.[3]

The pamphlet's anticapitalist, populist vision of a so-called "broad-based national defense state" (*kōgi kokubō kokka*) appealed to the plight of the masses. It denounced liberal capitalism, especially its profit orientation, intense competition, wealth disparities, unemployment, and urban and rural poverty. The army called for a new economic structure to "further the well-being of the entire people and the development of the state, based on the moral economic view reflected in the founding ideals of the state," as well as provide stability to the masses and

3. Takahashi Masae, ed., *Gendaishi shiryō*, vol. 5, 268.

improve the plight of workers. It also called for measures to revive agriculture and mountain and fishing villages in response to the deepening agricultural crisis as a result of the world depression. Suzuki Teiichi later recalled that he was influenced by left-wing ideas promoted by groups such as the Socialist Party and Social Masses Party and prominent activists such as Nishio Suehiro and Asō Hisashi. Suzuki claims that previously he had produced a radical plan for reform that included the nationalization of heavy industry, especially of electric power, coal, and iron production, shipping, and major railway lines; limits on private property and land ownership; and elimination of the peerage system.[4]

Military technocrats fared only slightly better than the radical officers in promoting their plans to reform Japan. The total war strategy of the Control faction was sharply criticized by senior army leaders of the Imperial Way faction such as War Minister Araki Sadao and Vice Chief of the General Staff Mazaki Jinzaburō. Araki sought to rid the army of technocrats by transferring and demoting key control officers such as Nagata Tetsuzan and Tōjō Hideki in 1933. Under the new war minister General Hayashi Senjūrō, however, Nagata was reinstated and became chief of the Military Affairs Bureau in 1934, only to be assassinated by a devoted follower of Mazaki the following year. The Control faction's army pamphlet managed to appease Imperial Way officers by calling for aid to the countryside, but it incurred the wrath of business and the political parties. The two major political parties, the Seiyūkai and Minseitō, denounced the army's plan to reorganize the economy and society for the purposes of national defense. They accused the army of dabbling in politics and posing as the official spokesmen on social and economic policy.[5]

Prioritizing External over Internal Reform

The distinct political agendas and strategies of the two groups were reflected in their different strategies to link the Manchurian invasion to domestic reform. At issue were the questions of the sequencing and prioritization of internal and external reforms. Despite their differences, however, both groups shared the basic conviction that capitalism, as they knew it, was immoral and detrimental to the long-term interests of Japan. Their mutual antipathy toward the status quo provided a common basis on which to devise an alternative type of state planned system.

4. Suzuki, *Suzuki Teiichishi danwa sokkiroku*, 339–342.
5. Ōtani Keijirō, *Gunbatsu: Ni-ni-roku jiken kara haisen made* (Tokyo: Tosho shuppansha, 1971), 127.

Collaboration between radical officers and military technocrats was difficult because each side embraced different conceptions of reform. Radical officers and their right-wing supporters understood "reform" in the cultural terms of a spiritual reunion of emperor and people. Military technocrats, in contrast, defined reform in the political-economic terms of a national defense state, a controlled economy, mechanization of military forces, and the creation of a military-civilian planning unit, or "economic general staff." Whereas the radical right promoted Japan's common cultural, ethnic ties with Asia as the basis for joint action against the West, technocrats sought a geopolitical, economic, and technological basis for the establishment of an autarkic Asian sphere led by Japan.

Not only did the two groups pursue different strategies, they often viewed each other's plans with mistrust and derision. The radical right counted technocrats among those who were the primary obstacles between the emperor and the people. Given the right wing's antibureaucratic, antiurban, and anti–big business stance, they viewed technocrats with particular hostility. They castigated reform bureaucrats as opportunists who skillfully took advantage of the collapse of political parties to "ride on the backs of the military" in order to become major political players. Control officers, in turn, accused young officers of being irresponsible and a threat to military prerogatives. Civilian technocrats such as Rōyama described the young officers as mere "restorationists" without concrete plans and "nationalists" without a coherent ideology.[6]

Despite their differences, the two groups shared a common fascist vision. Both were highly disillusioned with Japan's liberal capitalist order and the principles of private property and self-interest on which it was based. They sought to go beyond the materialism of capitalism and socialism and eliminate the "evils" of capitalism, particularly the "monopolization of capital" by big business and its short-sighted focus on individual profit. Most were not prepared to eliminate the constitution or, initially, the labor unions and political parties. They preferred to co-opt these organizations in ways to mobilize public support for state goals. These groups envisioned an authoritarian state and a holistic, organic society in place of what they perceived as the atomistic, conflict-ridden liberal capitalist society. They exalted public over private interests and the primacy of the national community over the individual. Moreover, they believed that the true freedom of the individual could only be achieved within the community—interpreted as the "national community" (*kokumin kyōdōtai*), or "national defense state." Finally, both groups embraced war and Japanese expansion in Asia as legitimate means to achieve their goals. They shared the conviction that Japan should liberate Asia from Western imperialism and revive its eastern culture and spirit.

---

6. Fletcher, *The Search for a New Order*, 56–57.

Young officers and army technocrats viewed the resolution of the Manchurian problem and the reform of Japan as closely related. In the weeks preceding the invasion of Manchuria, control officers in the Kwantung Army kept in close contact with radical officers. Their discussions centered on the issue of whether domestic reform should precede external expansion or vice versa. Both Ishiwara and Itagaki were concerned about anti-Japanese sentiment in Manchuria. After the Russo-Japanese War, Japan had pursued its economic and defense interests in Manchuria through railway leaseholds, its semipublic South Manchuria Railway Company, and the Kwantung Army, which was created to defend the railway. By the late 1920s, however, antipathy toward the Japanese had become acute and Japanese settlers had requested financial aid and political support from the Japanese government.

Kwantung Army leaders viewed the situation in Manchuria as critical but also offering a way forward. They argued against postponing a military offensive in Manchuria in the interests of domestic reform. In his "Personal View of the Manchuria and Mongolia Problem" of May 1931, Ishiwara pointed out the difficult challenges involved in domestic reform. Achieving political stability in Japan required time, and even if the political situation stabilized, there was still the task of reforming the economy. Detailed and timely economic planning was needed and domestic reforms could not be implemented until the economy deteriorated to a certain crisis point.[7] Ishiwara argued that war was an integral part of state reform. It was better to reform Japan under a wartime system because war would revive the economy, and under a military state of emergency, reforms could be implemented more easily. Hence Japan should first embark on external expansion and then reform Japan's political and economic system.

Radical officers argued that internal reform should precede external expansion. In one meeting in Tokyo in the summer of 1931, Kwantung Army officers discussed the problem with members of the right-wing Cherry Blossom Society, Lieutenant Colonel Hashimoto Kingorō and Colonel Nemoto Hiroshi. Both had been planning the abortive March and October incidents to replace the Minseitō cabinet of Wakatsuki Reijirō with a reformist one. Hashimoto proposed a coup d'état to overthrow Japan's party government before Japan took military action on the continent. Nemoto expressed concern over whether there would be sufficient support from the Wakatsuki cabinet for direct action in Manchuria. He suggested that the cabinet be toppled before Japan embarked on any exploits

---

7. "Manmō mondai shiken" (May 1931), in Nihon kokusai seiji gakkai taiheiyō sensō gen'in kenkyūbu, ed., *Taiheiyō sensō e no michi, vol. 8, Shiryō bekkan* (Tokyo: Asahi shinbunsha, 1962–1963), 101.

abroad.[8] At issue was the ideological basis of control and the type of system to be established in Manchuria. Above all, these officers were concerned that if the military expanded abroad before reforming Japan's corrupt capitalist system, their actions would result in the export and propagation of the very system that they were trying to reform at home. Therefore, Japan's expansion into Manchuria would be essentially no different from the imperialism of the Western powers and, as one officer delicately put it, "the military declined to be the palanquin carrier of capitalism."[9]

The Kwantung Army unilaterally concluded the debate by directly occupying Manchuria on September 18, 1931. On September 22, the decision was made to establish a "paradise of the various ethnic people of Manchuria and Mongolia" (*Man-Mō kaku minzoku no rakudo*) under some type of Chinese dynastic authority.[10] In early October of that year, army headquarters in Tokyo ordered the Kwantung Army to establish an independent state. Kwantung Army leaders were sensitive to the concerns raised by the radical right and sought to unite the two different conceptions of state reform. The challenge was to channel the powerful force of ethnic nationalism and the restorationist spirit into economic and political reform and reconcile the romantic, spiritual ideals of the radical right with technocratic visions of a national defense state.

Justifying Manchukuo

Studies of Japanese imperialism have grappled with the question of why the Manchurian venture differed from Japan's colonial ventures in Korea and Taiwan. Why were leaders keen to present Manchukuo as an independent state rather than as a formal colony?[11] It is true that in the post–World War I era of "self-determination" and "Open Door," nineteenth-century style colonies, like nineteenth-century wars, were considered an anachronism. There were also pragmatic reasons for army leaders to establish an independent state. The army required the cooperation of local groups and took up their pan-Asianist

---

8. Ōtani quotes Ikeda Sumihisa, see Ōtani, *Gunbatsu,* 169.

9. Ibid.

10. Katakura Tadashi and Furumi Tadayuki, *Zasetsushita risō koku: Manshukoku kōbō no shinsō* (Tokyo: Gendai bukkusu, 1967), 47, 58; Komagome Takeshi, "'Manshūkoku' ni okeru jukyō no isō: Daidō, ōdō, kōdō," *Shisō* 851 (July 1994), 60.

11. Yamamuro Shinichi, *Manchuria under Japanese Dominion,* trans. Joshua A. Fogel (Philadelphia: University of Pennsylvania Press, 2006), esp. Introduction; Komagome Takeshi, *Shokuminchi teikoku Nihon no bunka tōgō* (Tokyo: Iwanami shoten, 1986), chapter 5; Prasenjit Duara, *Sovereignty and Authenticity: Manchukuo and the East Asian Modern* (Oxford: Rowman & Littlefield, 2003), Introduction and 61–65.

cause as a price for their assistance. Eventually some Japanese leaders like Ishiwara became enthusiastic proponents of an Asian brotherhood and an "East Asian League." But another motive for establishing an independent state can be found in the reformist ideals of its founders. From the start, Kwantung Army officers viewed Manchukuo as the antithesis of liberal capitalist Japan and as one component of domestic reform. Above all, they wanted Manchukuo to be independent from Japan's ruling class and especially its corrupt capitalists and politicians.

The Manchurian occupation signified a novel approach toward empire. It marked a new phase of Japanese imperialism that was distinct from Meiji imperialism. Meiji leaders had conceived of Taiwan and Korea as colonial showcases that represented the outward face of Japan's modernization. They believed that in order for Japan to stand shoulder to shoulder with the West and eliminate the unequal treaties, Japan needed not only a constitutional government and capitalist economy but also a colonial empire. Japan practiced Western-style "gunboat diplomacy" toward Korea in the late nineteenth century before formally annexing it in 1910 and, later, "cooperative diplomacy" in Manchuria and China to secure its interests. But by the early 1930s, Japanese technocrats rejected Western-style imperialism. They justified the Manchurian occupation not in the Western liberal terms of profits, the search for markets, and a civilizing mission, but in terms of geopolitics and ethnic-chauvinist visions of pan-Asianism.

The diplomat Matsuoka Yōsuke officially justified the seizure of Manchuria with the geopolitical argument that it represented Japan's "lifeline." Proponents of the "movement to establish the lifeline" (*seimeisen kakuritsu undō*) argued that Manchuria's geographical proximity to Japan, strategic border with Russia and Korea, and abundant natural resources made it a critical area for Japanese national defense and total war mobilization. They claimed that Japan had a "right to exist" (*seizonken*) and expand at the expense of its Asian neighbors.[12] Through its Manchurian lifeline, "Great Japan" (*dai Nihon*) would become a superpower (*chōdaikokka*) and "dance upon the high class stage" with England, America, and Russia.[13] Geopolitical theories drew on the Social Darwinian notion of the state as a biological organism that was engaged in a struggle for survival of the fittest. Expanding on this idea, geopoliticians argued that certain states required living space to grow. The German geopolitician Karl Haushofer, who was a

12. Katō Shinkichi, "Manshū ni okeru minzoku kyōwa no mondai—toku ni Nihonjin wa ikanishite ikieruka ni tsuite," in Kantōgun, ed., *Manshū jihen jisshi* (Tokyo: Nittō shōin, 1932), 413.

13. Koyama Sadatomo, "Seimeisen kakuritsu undō," *Manshū hyōron* (October 1931), 23; Tachibana Shiraki, "Manshū jihen to fuashizumu," *Manshū hyōron* (November 1931), 8–9.

great admirer of Japan, envisioned the emergence of several semiautarkic "pan-regions" centered on Germany, the United States, Japan, and possibly Russia.[14]

Geopolitical concepts were promoted at the highest levels of government, especially by influential technocrats such as Matsuoka Yōsuke, Rōyama Masamichi, and Mōri Hideoto. In his own writings since the late 1920s, Rōyama, who advised the Kwantung Army, had argued for Japan's special relationship with Manchuria based on the "organic economic relationship" between the two countries.[15] He used geopolitical concepts to develop his "Asian Monroe Doctrine" for a Japan-centered regional bloc. Mōri, who would later construct the ideological framework for the New Order, drew on geopolitical concepts to justify Japanese expansion in Asia. He used numerous birth metaphors in his references to East Asia as a "living body," and to the "labor pains of the expansive power of the Japanese continent" and its "conception" or "fertilization."[16]

Japanese technocrats found in the concept of "living space" a means to critique the Western-dominated world order. Since the mid-1920s Japanese geopoliticians had called on the Asian people to develop a geopolitical awareness in order to fight against the "White Peril"—an allusion to Western denunciations of Japan as the "Yellow Peril" and American anti-Asian immigration policies.[17] They criticized the unequal division of the world on the grounds that an ethnic people who lack space will degenerate. In the late 1930s, technocrats argued that "have-not" countries like Germany and Japan had the right to acquire colonies to support their expanding populations. Geopolitics appealed to army leaders because it enabled them to present Japan as the defender of Asia against the West while simultaneously pursuing its strategic military interests in the region. Moreover, it provided a pseudo-scientific justification for Japan to secure its own regional trading bloc amidst the worldwide trend toward "monopolistic trading blocs" and the dramatic shrinkage in volume of international trade during the depression.

Official pan-Asianist visions of Manchukuo were incorporated as a result of practical considerations surrounding the founding of the new state. The Kwantung Army was not powerful enough to unilaterally dictate policy. It lacked sufficient funds, institutional and logistical support, and political and economic

14. See Derwent Whittlesey, "Haushofer: The Geopoliticians," in Edward Mead Earle, ed., *Makers of Modern Strategy* (New York: Atheneum, 1966).

15. Rōyama Masamichi, *Japan's Position in Manchuria,* (The Japan Council, Institute of Pacific Relations, 1929), 79.

16. Kamakura Ichirō, Tōa kyōseitai kensetsu no shojōken," *Kaibō jidai* (October 1938), 26.

17. Hatano Sumio, "'Tōa shinchitsujo' to chiseigaku," Miwa Kimitada, ed., *Nihon no 1930 nendai* (Tokyo: Sōryūsha, 1980), 17.

expertise to build the new state.[18] Army leaders relied on Mantetsu's political clout and expertise. They sought the counsel and support of its researchers, Japanese activists, merchants, regional bureaucrats, and the local Chinese elite for their knowledge of Manchurian society. Many of the Japanese who advised the army could be classified as part of the dispossessed strata of Japanese society that came to Manchuria to seek a brighter future. Many were hard hit by the depression and harbored resentment against both the privileged elites in Japan and profitable Chinese merchants in Manchuria. They included young officers in the Kwantung Army, anticommunist Soviet experts, Marxist sinologists, members of the pan-Asianist Manchurian Youth League and Great Majestic Peak Association (Daiyūhōkai), Japanese settlers, and adventurers. The army was also assisted by Japanese-speaking Chinese regional leaders and landowners who curried favor with the army in order to protect their own interests under the new regime.

Army leaders and their supporters shared a mutual aversion to liberal capitalism, but possessed different ideas about the new state. A number of left- and right-wing leaders were highly critical of the official view of Manchuria as Japan's lifeline. They promoted their own plans to create a utopian, independent state and develop Asia. One influential right-wing advisor who eventually clashed with army leaders was the Mantetsu researcher Kasagi Yoshiaki. A former classmate and colleague of the pan-Asianist Ōkawa Shūmei at the company's East Asian Research Bureau in Tokyo, Kasagi transferred to Manchuria and formed his own right-wing group, the Great Majestic Peak Association.[19] He recruited young activists at Mantetsu, including graduates from his alma mater the Fifth Higher School in Kumamoto. Kasagi's group denounced the "Manchurian lifeline" view as "no more than an expression of Japanese egoism" and insisted that "Manchuria is Manchuria" and not an appendage of Japan. His criticism of government policy later got him into trouble with the army and he became the target of several assassination attempts.[20]

On the left was Tachibana Shiraki, a respected Mantetsu sinologist and a close friend and former classmate of Kasagi. Tachibana employed a Marxist framework in his analysis of Chinese society and championed nationalist revolution to liberate the Chinese peasantry from despotic feudal rule. Tachibana was at odds with the main thrust of Kwantung Army policy. In the *Manchurian Review,* of which he was editor, Tachibana criticized military policy by drawing parallels

---

18. Yamaguchi Jūji, *Kieta teikoku Manshū* (Tokyo: Mainichi shinbunsha, 1967), 102, 219.

19. Imura Tetsuo, ed., *Mantetsu chōsabu: Kankeisha no shōgen;* Kusayanagi Daizō, *Jitsuroku Mantetsu chōsabu* (Tokyo: Asahi shinbunsha, 1979), 90.

20. Kusayanagi, *Jitsuroku Mantetsu,* 91–92.

between certain aspects of military thinking and fascism.[21] He pointed out the fascist tendencies in the military's imperialist lifeline policy to realize its dream of "Great Japan." For Tachibana, fascism was no more than the "petty bourgeois ideology of the small farmer and laborer." Fascism rejected monopoly capitalism and communism and embraced some kind of national socialist dictatorship and state capitalism, while asserting its own supremacy over other ethnic peoples and states. Tachibana believed that the military's policy was fascist because it rejected parliamentary politics and class struggle and espoused the anticapitalist rhetoric of small farmers and laborers, while advancing the interests of monopoly capitalists.

Idealists such as Kasagi and Tachibana maintained an ambivalent stance toward army policy. Despite their aversion to military fascism, both men believed that collaboration was the best way to promote the interests of the Manchurian people. As employees of Mantetsu, headquartered in Dairen (Dalian), they adhered to the "Dairen ideology" that the company was not simply a vehicle for Japanese imperialism. Its mission was to fight against the imperialism of the white man and to create an East Asian cultural sphere through projects such as the building of schools, hospitals, and a railway system.[22]

Indigenous Concepts of State and Economy

Establishing an independent Manchurian state required resolving a set of contradictions in Japanese policy. How to make a case for Manchurian autonomy from China when the majority of the population was Chinese? How to justify Japanese rule in which a small group of Japanese in the new capital of Shinkyō (Changchun) ruled over the Chinese, Manchurian, Mongolian, and Korean masses in the hinterland? The army needed to devise principles of rule that possessed legitimacy in the eyes of the Chinese local elite. It needed to find a way to appeal to Chinese cultural and political tradition while building a national defense state. Manchuria had various ethnic groups and no state. How to best promote a collective form of ethnic nationalist identity centered on the new state? The army sought political principles to portray Manchukuo as both an ethnic nation and a modern Chinese state. Based on the ideas of Japanese and Chinese advisors, it adopted as its mandate the construction of a state based on the Chinese political concept of ōdō ("the way of the monarch" or "kingly way") under the banner of minzoku kyōwa ("ethnic harmony"). In the process, Kwantung Army officers

21. Tachibana, "Manshū jihen to fuashizumu," 2–3.
22. Kusayanagi, *Jitsuroku Mantetsu,* 130.

developed their own fascist vision of an authoritarian, organic, national defense state and planned economy led by the army's vanguard party, the Concordia Association (Kyōwakai), which promoted the ethnic nationalist principles of kingly way and ethnic harmony.

Army leaders found in the concept of kingly way a means to justify the creation of an independent state from China and the restoration of the Qing emperor Pu Yi. The concept was brought to the army's attention by Tachibana. Through his writings and discussions with Kwantung Army leaders, Tachibana, together with his colleague Noda Ranzō, convinced army leaders that these principles of government possessed a sound theoretical basis and would be accepted readily by the Chinese.[23] According to Tachibana, the concept of kingly way had been promoted by Mencius and drew on Buddhist notions of the sage-king who rules the earth based on the mandate of heaven. *Ōdō* depicted a monarch who bears responsibility to both heaven and the realm to exercise virtuous rule by respecting the people's will; its opposite was *hadō,* or rule by force. When the people's trust was lost, discord arose in the realm and the monarch lost heaven's mandate to rule. Kingly way government aimed at the realization of the Buddhist utopian world of *daidō* (Great Moral Principle) based on three conditions: securing the livelihood of the people, developing the riches of the realm for public use, and exerting one's labor for society's benefit.[24] "The Great Moral Principle" represented a timeless ideal against which the present government could be assessed.

Sinologists closely associated kingly way with the concept of "secure the borders and pacify the people" (*hokyō anmin*), which had been advocated by Chinese leaders in Manchuria prior to the Japanese invasion. "Secure the borders and pacify the people" argued for the separation of Manchuria from China and an end to the disruptive military campaigns into China by local warlords such as Zhang Zuolin and Zhang Xueliang. The concept was associated with the so-called "civilian faction" (bunchi-ha) in Mukden led by Wang Yongjiang, Yu Chonghan, and Yuan Jinkai. This faction had sought to isolate Manchuria from the civil wars in China and create a sound government and prosperous economy in Manchuria. As a movement promoted by landlords it was essentially conservative. Tachibana advanced a revisionist interpretation that the Japanese were joining together with the Chinese landlords and masses to liberate Manchuria from warlord rule and capitalists.[25] *Ōdō* supported the ideal of the traditional

23. Hashikawa Bunzō, "Tōa shinchitsujo no shinwa," in Hashikawa Bunzō and Matsumoto Sannosuke, eds., *Kindai Nihon seiji shisōshi,* vol. 2 (Tokyo: Yūhikaku, 1970), 353–354; Komagome, "'Manshūkoku' ni okeru jūkyō no isō: Daidō, ōdō, kōdō," 62.

24. Tachibana Shiraki, "Ōdō seiji," *Manshū hyōron* (May 1932), 4.

25. On Tachibana's collaboration, see Louise Young, *Japan's Total Empire: Manchuria and the Culture of Wartime Imperialism* (Berkeley: University of California Press, 1998), 282–291; Hirano

Chinese village community: autonomous, self-governing, and dominated by the landlord. This vision of an autonomous village community, which had been promoted by Japanese right-wing agrarianists such as Tachibana Kōsaburō and Gondo Seikei, had also appealed to radical young officers.

The concept of kingly way was subsequently adopted as the independent movements by Chinese leaders got underway. To support this movement, Tachibana and Kasagi founded the Self Government Guidance Unit (Jichi shidōbu) to promote autonomous local rule. Tachibana linked the idea of the sage-ruler and the autonomous village community together by arguing that although the self-supporting local society based on the idea of "secure the borders and pacify the people" was the ideal, in times of hardship it was the duty of the sage-king, Pu Yi, to provide leadership and restore the powers of self-government in Manchuria. Kingly way was a convenient slogan for army leaders because they could present themselves as Pu Yi's advisors who would protect traditional Chinese society against the utilitarian interests of the West (hadō). As for the notion of liberation of the Chinese masses, the army was primarily interested in making the Chinese landlord elite identify with the new state and not in awakening the masses for revolution.

Having justified the goal, the means needed to be justified. Under what principle could the Japanese undertake the task of promoting kingly way? In its first official statements concerning the new government, army leaders stated as its goal the construction of a "paradise of the various ethnic people of Manchuria and Mongolia" (Man-Mō kaku minzoku no rakudo) or "paradise of co-prosperity and coexistence" (kyōson kyōei no rakudo).[26] The army adopted the slogan of ethnic harmony among the people in Manchuria—the Chinese, Manchus, Mongols, Japanese, and Koreans. There was considerable support among Japanese civilians in Manchuria for this idea, especially among the right wing. Members of the Manchurian Youth League, whose interests were most directly threatened by Chinese nationalism, had actively promoted it before the incident with the aim to neutralize an increasingly hostile anti-Japanese political climate. Appealing to Confucian notions of a hierarchical society that they had used in their previous slogans of "co-prosperity and coexistence" and "Japanese-Chinese harmony" (Nikka wagō), the army promoted ethnic harmony to foster peaceful coexistence of the various ethnic peoples within a segmented society. Moreover, they hoped

Ken'ichirō, "The Japanese in Manchuria, 1906–1931: A Study of the Historical Background of Manchukuo" (Ph.D. diss., Harvard University, 1983), 399–415.

26. The first term is used in "Manmō mondai kaiketsu sakuan" (September 1931); the second term is used in "Manmō mondai kaiketsku an" (October 4, 1931). See Katakura and Furumi, Zasetsushita risō koku: Manshūkoku kōbō no shinsō, 47, 58.

that ethnic harmony would not only justify a role for the Japanese, but also help ameliorate ethnic tensions between the Koreans and Chinese in Manchuria.

The Kingly Way State

In his penetrating study of Manchukuo, the Japanese historian Yamamuro Shin-ichi expressed the complexity and inner tensions and contradictions of the Manchurian state by portraying it as a "Chimera."[27] This imaginary beast of Greek mythology possessed the head of a lion, the body of a goat, and the tail of a serpent. In pursuing the analogy, Yamamuro likened the head to the Kwantung Army, the body to the emperor system, and the tail to the Chinese emperor and state. In the early phase of state building, however, the leaders of the new state had envisioned something more unified and holistic. Manchukuo's leaders viewed the state as an organism, in which the different organs were vital to the functioning of the whole. The Kwantung Army's strategic planning unit served as the brain that devised and coordinated the policies of the new state. The Concordia Association functioned as the heart, whose pan-Asianist ideals circulated through the state and infused it with idealism and spirit. The state bureaucracy comprised the flesh, bones, and muscle that gave it form and strength. Despite claims of a Manchurian lifeline, Manchukuo was designed not merely as a supply line to or appendage of Japan, but as an independent, growing "fetus" in its own right, although eventually it was subsumed by the "mother" country of Japan.

Manchukuo's founding ideals were expressed organizationally and transformed into a functioning reality in the new structures of the state. In contrast to Korea and Taiwan, Manchurian rule did not take the form of a hierarchically structured, top-down chain of commands by Japanese governors-general. Manchukuo was designed as a totalist state in which the totality or whole took precedence over the individual parts. It was governed via concentrically arranged spheres of power in which real authority emanated from the inner core.

In the "outer layer," the new state took the form of a "democratic constitutional republic" (*minshū rikken kyōwa sei*) modeled on the Nationalist Government's three-branch system composed of an executive wing, the Government Affairs Board (Kokumuin) headed by the prime minister; a Legislative Board (Rippōin); and Supervisory Board (Kansatsuin). The formal head of state was the former Qing emperor, Pu Yi, who became chief executive. Pu Yi was advised by the Councilors Office (Sangifu), which corresponded to Japan's Privy Council. The executive wing oversaw the General Affairs Agency (Sōmuchō) and seven

---

27. Yamamuro, *Manchuria under Japanese Dominion*, 6–8.

departments for civilian affairs, diplomacy, military affairs, finance, industry, transportation, and justice. As a rule, the Chinese assumed the posts of prime minister, department head (minister), and branch head. In order to maintain the façade of independence, as well as to balance power among the former local Chinese elite, the Kwantung Army appointed the leaders of the Chinese "independence" movements to head the various offices. As a "democratic" republic based on the will of the people of Manchuria, it was necessary to have a legislative branch, even if it had no real function to speak of.[28]

Beneath this external layer of formal Chinese rule, was the "inner layer" of Japanese administration. Japanese bureaucrats were concentrated within the General Affairs Agency and the General Affairs Section (Sōmushi) within each of the seven departments. The General Affairs Agency handled all important and confidential matters related to personnel, the budget, and resources. The secretariat, or General Affairs Section, within each of the seven departments, as well as in the regional branch, essentially mirrored the functions of the General Affairs Agency at a lower level.[29] The General Affairs Agency and General Affairs Sections have been compared to Japan's cabinet secretariat (Naikaku shoki-kyoku and Daijin kanbō).[30] The most powerful posts within the new state were those of chief, deputy chief, and section head within the General Affairs Agency, and deputy chief (vice minister) and General Affairs Section head within each department.[31]

The policymaking process reflected the discrepancy between the outer and inner layers of rule. Formal deliberation of national policy took place at the weekly meetings of the Government Affairs Board (Kokumuinkaigi) attended by the Chinese prime minister and department heads. The agenda for these meetings, however, was discussed ahead of time within the General Affairs Agency and its decisions were forwarded to the emperor and the Councilors Office. In reality, actual policy decisions were made not at the Government Affairs Board meetings, but at the informal weekly meetings of senior Japanese bureaucrats who prepared the agenda. These so-called "Wednesday meetings" were led by the chief of the General Affairs Agency and attended by the Japanese deputy heads

28. Yamamuro Shinichi, "'Manshūkoku' tōchi kateiron," in Yamamoto Yūzō, ed., "Manshūkoku" no kenkyū (Tokyo: Ryokuin shobō, 1993), 89–91.

29. Yoshida Yutaka, "Gunji shihai: Manshū jihen ki," in Asada Kyōji and Kobayashi Hideo, eds. Nihon teikokushugi no Manshū shihai: Jūgonen sensōki o chūshin ni (Tokyo: Jichōsha, 1986), 131.

30. See Rōyama, "Seiji," in Man-Mō jijō sōran (Tokyo: Kaizōsha, 1932), 110.

31. Fixed ratios of Japanese to Manchurians were determined for each department, respectively: 7:3 within the General Affairs Agency; 6:4 within the ministries of finance and industry; 3:7 within the ministries of civilian affairs, education, and diplomacy; 2:8 in the regional branch.

of the various departments and agencies.[32] Bureaucrats described this practice as bureaucratic rule based on the principle of "centralization of power within the General Affairs Agency" (*Sōmuchō chūshinshugi*).

At the core of the new structure was the Kwantung Army's own control apparatus through which it conveyed its will via "internal guidance" (*naimen shidō*) over Japanese bureaucrats. The name of the army's key unit for strategy and planning changed numerous times, from Third Division (Sanbō dai-sanka) at the time of the invasion, to Control Unit (Tōchibu) in December 1931, Special Affairs Unit (Tokumubu) in February 1932, back to Third Division, and later Fourth Division.[33] With each organizational change, the army sought to increase its leadership and surveillance powers over the state. In the early years, the Special Affairs Unit worked closely with Mantetsu's Economic Research Association in drafting the economic control policies for the new state, and together these two organizations functioned as a provisional Economic General Staff.[34] In the later years, planning and research were gradually assumed by Japanese bureaucrats. The army also exercised authority via the Councilors Office and its Secretariat (Hishokyoku). The Councilors Office was the highest advisory organ to the Chief Executive and, in practice, assumed the functions of the legislative branch.[35] It offered opinions and gave final consent to all important matters concerning laws, the budget, defense, treaties, and official proclamations. Its committee agendas as well as final decisions were transmitted via the head of the Secretariat. The army planned to exert authority through Japanese officials appointed to the Councilors Office and the post of Secretariat head.[36] Through an advisory system (*shimonsei*) established in January 1932 the Kwantung Army also planned to exert influence through the dispatch of both formal and informal advisors to the various departments in the central and regional governments.[37]

32. Participants included the vice ministers of each department, chief of General Affairs Section in each department, head of the Legal Bureau (*Hōseikyoku*), the deputy chief and section heads of the General Affairs Agency, the Kwantung Army chief of staff, the chief military advisor to the defense department, and the head of the Concordia Society.

33. Hara Akira, "1930 nendai no Manshū keizai tōsei seisaku," in Manshūshi kenkyūkai, ed., *Nihon teikokushugika no Manshū* (Tokyo: Ochanomizu shobō, 1972), 9; Mutō Tomio, *Watakushi to Manshūkoku* (Tokyo: Bungei shunjū, 1988), 113.

34. In 1932 army and Mantetsu planners began to advocate the establishment of an Economic General Staff both in Japan and Manchukuo. See Minami Manshū tetsudō kabushikigaisha keizai chōsakai, *Manshū keizai tōsei hōsaku* 1, no. 1 (*gokuhitsu*), 24, 72.

35. Manshūkokushi hensan kankōkai, ed., *Manshūkokushi: Kakuron* (Tokyo: Manmō dōhō engokai, 1971), 5.

36. Originally three Japanese officials were to be appointed to the Councilors Office. Candidates for the post included Ōkawa Shūmei and Itō Bunkichi. In the end only one councillor was appointed.

37. Yamamuro, "'Manshūkoku' tōchi kateiron," 103; Furumi Tadayuki, *Wasureenu Manshūkoku* (Tokyo: Keizai ōraisha, 1978), 48.

The "conscience" of the state, which was to provide the "moral halo" for Japanese imperialism, was the official mass propaganda organization, the Concordia Association.[38] Manchukuo was not conceived as simply a military dictatorship geared toward total war, but as a new type of state and economy founded on pan-Asianist ideals.[39] Ishiwara, who became an enthusiastic supporter of the state's pan-Asianist goals, conceived of the Concordia Association as a vanguard party that could serve as a unifying political force to directly mobilize the masses and guide government policy. Ishiwara and members of the Manchurian Youth League such as Yamaguchi Jūji had first proposed the creation of a Concordia Party (Kyōwatō) to represent a state party (*ikkoku ittō*) centered on the youth of the various ethnic peoples.[40] The name was changed to Concordia Association, however, due to the opposition of the incoming Kwantung Army leader Koiso who wanted to avoid any association with "party" politics. The official platform of the Concordia Association proclaimed that Manchukuo was the antithesis of the governments of Japan, England, China, and Soviet Russia, and adopted as its platform the rejection of the "capitalist monopoly," "parliamentary dictatorship," "military dictatorship," and "communist dictatorship" of these countries, respectively.[41] It assumed the role of safeguarding the ideals of the new state, particularly vis-à-vis Japan, as it announced in its statement of principles:

> [For] Japan to try to fight against the Chinese people in Manchukuo under its political authority and support from now on would mean to imitate the Chinese military cliques. [Japan] certainly cannot claim magnanimity in being able to enter the battle ring against the white man as king of East Asia. We, who broke down the fundamental barriers to the friendship and goodwill of the various East Asian peoples, demand to realize the ethnic ideals through harmonious and just struggle first in Manchukuo, and thereby take the first step here toward the harmony of all nations and the goodwill and friendship of the various East Asian peoples.[42]

---

38. I adopt the term *moral halo* from Speier, "Ludendorff: The German Concept of Total War," 318.

39. Speier's distinction between Ludendorff's military dictatorship and Hitler's fascist dictatorship is useful, see ibid., 307.

40. Furumi, *Wasureenu Manshūkoku*, 139.

41. Itō Takeo, *Life along the Manchurian Railway: The Memoirs of Itō Takeo*, trans. Joshua A. Fogel (New York: M.E. Sharpe, 1988), 144; Koyama, *Manshū kyōwakai no hattatsu* (Tokyo: Chūō kōronsha, 1941), 33.

42. Koyama, *Manshū kyōwakai no hattatsu*, 38.

The Concordia Association took the position that "Manchukuo does not rely on the political dominance of the Japanese state; it is an independent state of ethnic harmony in which the Japanese people participate."[43]

A competing mass organization was also advanced by Kasagi to promote his pan-Asian vision of "a political system based on the ideals of the Eastern philosophers" and regional autonomy.[44] Kasagi came close to realizing his vision of an executive-level National Affairs Board (Shiseiin). This organization, which would consist of private schools to "promote the development of Asia" (Kōajuku) and a research unit to study the founding ideals of the state, would link the regions and center via a training center (*kunrensho*) for the explicit purpose of producing pan-Asianists—not "able bureaucrat types." A modified version was adopted in the form of a National Affairs Bureau (Shiseikyoku) within the Department for Civilian Affairs, which included a training center, research unit, and a propaganda arm to build support for the new state and prepare propaganda materials for foreign consumption. Radical officers and right-wing ideologues, including Kasagi, Hanaya Tadashi, Imada Shintarō, and Ōkawa, recruited around eighty right-wing students in Japan to join the training center. However, they adopted a critical stance toward what they referred to as "insincere bureaucratism" (*fushinsetsu na kanryōshugi*), which eventually clashed with official conceptions of the new state. Within a few months the National Affairs Bureau was abolished. Kasagi returned to Japan and continued to promote his pan-Asianist views at right-wing nationalist societies such as the Yūzonsha, founded by Ōkawa and Kita in 1919, and the Kōchisha, created by Ōkawa in 1923.[45]

The Kingly Way Economy

Kwantung Army leaders were keen on expressing Manchurian "independence" from Japan in its plans for the economy. From the start they embraced the policy of "excluding zaibatsu" and "excluding capitalists" (*zaibatsu hairu bekarazu, shihonka hairu bekarazu*). They aimed to build an independent, heavy industrial base to support their total war policy. Manchuria's economic program reflected both the army's technocratic vision of a national defense state and the utopian, anticapitalist, petite bourgeois mentality of its supporters. One official history of Manchukuo defined its twin goals as the creation of a "national defense

43. Ibid., 39.

44. Manshūkokushi hensan kankōkai, ed., *Manshūkokushi*, vol. 1, *Sōron* (Tokyo: Manmō dōhō engokai, 1971) 243–249; Furumi, *Wasureenu Manshūkoku*, 37–38.

45. Manshūkokushi hensan kankōkai, ed., *Manshūkokushi: Sōron*, 247–248; Itō, *Life along the Manchurian Railway*, 144–145; Imura, *Mantetsu chōsabu*, 360, 746.

economy" (*kokumin keizai*) for the purposes of attaining the military's strategic defense goals and an "economy for the well-being of the people" (*anmin keizai*) that realizes the principles of kingly way and secure the borders and pacifying the people.[46] The state's economic charter, "Outline of the Construction of Manchukuo's Economy" of March 1933 (*Manshūkoku keizai kensetsu yōkō*), described the state's "great mission" as the "completion of a new economic organization incomparable in the world, which is based on the eternal prosperity of the people and the state."[47]

In the outline, army leaders set out the principles for economic development. They declared that "in view of the evils of the uncontrolled capitalist economy, our economic construction aims at the attainment of health and vitality of the entire national economy by applying necessary state controls and making effective use of capital." The state promoted four basic policies: priority of public interests over the interests of one particular class; comprehensive development of the economy and resources through state control of important industries; encouragement of foreign capital and technology, especially from the advanced countries; and close economic cooperation between Japan and Manchuria.[48] The vehicle for economic development would be the state-run "special companies" (*tokushu kaisha*), which would manage all important industries related to defense and essential public services, while the remaining industries would be privately managed.[49]

The outline reflected a compromise between the idealistic, anticapitalist approach of Ishiwara's founding faction and the more pragmatic approach of incoming Kwantung Army leaders General Mutō Nobuyoshi and Lieutenant Colonel Koiso Kuniaki. While the initial research and drafts of the outline were made under Ishiwara's group, final drafts were completed under Mutō and Koiso. In the final version, the army's previous anticapitalist rhetoric to "keep out the zaibatsu" was toned down to an attack on the "monopolization of profits by one class"; foreign capital was welcomed under the policy of "open door" and "equal opportunity"; and a sphere for free enterprise was demarcated, if vaguely. The effect, however, was that the outline was roundly attacked by both sides. The business community and ministries in Japan denounced the proposed control measures as a barrier to investment. Ishiwara's supporters criticized the plan as a sell-out to mainland interests and, especially with regard to the policy to import

---

46. Manshūkoku seifu, ed., *Meiji hyakunen shi sōsho, vol. 91, Manshū kenkoku jūnenshi* (Tokyo: Hara shobō, 1969), 304.

47. Fujiwara Yutaka, *Manshūkoku tōsei keizai ron* (Tokyo: Nihon hyōronsha, 1942), 30.

48. Minami Manshū tetsudō kabushikigaisha keizai chōsakai, *Manshū keizai tōsei hōsaku*, 4.

49. Manshūkokushi hensan kankōkai, ed., *Manshūkokushi: Sōron*, 383.

foreign capital, as seriously compromising the anticapitalist views of the Concordia Association."[50]

Manchurian industrialization represented Japan's first attempt at fascist economic planning. In contrast to Fascist Italy and Nazi Germany, Japan possessed its own experimental base to try out fascist economic policies. Fascist economies are characterized by their military-strategic orientation, anticapitalist and anticommunist stance, promotion of nationalist goals, and extensive state intervention.[51] They are distinct from liberal capitalist economies in being driven by politics rather than the market. Some scholars reject the notion of a distinct type of fascist economy and view the economies of fascist regimes as no more than "an adjunct of fascist politics" or "capitalism with a cudgel."[52] The differing views about fascist economies derive from the discrepancy between their visions and reality. Fascists promoted anticapitalist, nationalist, military-strategic visions of an ethnic community-based economy while ultimately relying on and serving the interests of big business or "monopoly capital." In the case of wartime Japan and Manchukuo, after the initial stage of development, Japanese leaders relied heavily on big business while appealing to anticapitalist notions of "separating capital and management" and building a national defense economy. Their reliance on "monopoly capital," however, did not reflect a cynical ploy to deceive the masses. They believed that capitalism was being superseded by a new type of managerialism, in which technocrats, not capitalists, directed industry and served state and ethnic nationalist goals. As we will see in the following chapter, reform bureaucrats sought to replace Manchuria's state capitalism with a new type of antiliberal managerialism directed by technocrats.

As in state building, Kwantung Army leaders improvised and developed Manchuria's economy with the materials at hand. Manchuria's economy was primarily agrarian and relatively poor in terms of industry, accumulated capital, technology, and a skilled labor force. Its undeveloped infrastructure, combined with the military's ambitious timetable for heavy industrialization, demanded some sort of state-directed program. The most relevant model for Manchurian development was the socialist model. The Soviet Union, which launched its first Five Year Plan in 1927, had faced similar types of challenges. The distinctive characteristics of the Soviet planned economy were its use of the state plan, the command, and a specialized planning apparatus. In such a system, the state

---

50. Yoshida, "Gunji shihai: Manshū jihen ki," 145; Koyama, *Manshū kyōwakai no hattatsu*, 39.

51. Bai Gao, *Economic Ideology and Japanese Industrial Policy: Developmentalism from 1931 to 1965* (Cambridge: Cambridge University Press, 1997), especially 11–12, 23–28; S. J. Woolf, "Did a Fascist Economy Exist?" in S. J. Woolf, ed., *The Nature of Fascism* (London: Weidenfeld: Nicolson, 1968).

52. Charles Maier, "Economics of Fascism and Nazism," *In Search of Stability* (New York: Cambridge University Press, 1987), especially 71, 116.

relied on various short- and long-term "plans" containing numerical targets for production, distribution, labor, investment, finances, and technological innovation. Implementation of these plans was compulsory and effected through top-down commands backed by incentives and penalties, including the use of force.[53] Plans were implemented and managed via a centralized planning apparatus that essentially replaced the market mechanism. The state bureaucracy, not the market, approved the entry and exit of firms; appointed managers; allocated labor, materials, and products; and determined prices.

Army leaders relied on Mantetsu economists knowledgeable about Soviet planning, as well as research on the controlled economy undertaken by Marxist economists in Manchuria and Japan.[54] The army's most influential economic advisor on Soviet planning was Miyazaki Masayoshi, a graduate of the University of Petersburg who had witnessed the Russian revolution.[55] Before the invasion Miyazaki had fostered close ties with senior Kwantung Army officers in Manchuria such as Ishiwara and Itagaki Seishirō. He helped the army establish the Economic Research Association (Keizai chōsakai) and recruited researchers from Mantetsu to join its staff. In the early years this research association, together with the army's Special Affairs Unit, served as the main economic advisory organ to the government.

Officials distinguished Manchuria's economy from both Japan's controlled economy and the Soviet Union's planned economy. They argued that Manchuria's economy belonged to the genre of "planned economy" not "controlled economy." In a controlled economy (tōsei keizai), piecemeal reforms are implemented by the state in order to address the contradictions arising out of the existing market-based capitalist system. Manchuria's economy was a planned economy because it represented a "new national economic system that attempts to regulate supply and demand in the process of [capital] reproduction of the national economy under an established comprehensive plan for the purpose of realizing the state's own goals."[56] At the same time, they distinguished Manchuria's "partially planned economy" (bubunteki keikaku keizai) from the Soviet Union's "total planned economy"

53. János Kornai, The Socialist System: The Political Economy of Communism (Princeton: Princeton University Press, 1992), chapter 7.
54. Mantetsu researchers could also draw on the extensive body of theoretical work on the controlled economy in Japan undertaken by Marxist economists such as Yamada Moritarō, Arisawa Hiromi, and Sakisaka Itsurō. See the multi-volume series edited by Sakisaka Itsurō: Nihon tōsei keizai zenshū, vols. 1–10 (Tokyo: Kaizōsha, 1933–1934) and Arisawa Hiromi, Sangyō dōin keikaku (Tokyo: Kaizōsha, 1934).
55. For a study of Miyazaki Masayoshi, see Kobayashi Hideo, "Nihon kabushikigaisha" o tsukutta otoko: Miyazaki Masayoshi no shōgai (Tokyo: Shōgakkan, 1995).
56. Manshūkoku seifu, ed., Meiji hyakunenshi sōsho, vol. 91, Manshū kenkoku jūnenshi (Tokyo: Hara shobō, 1969), 321.

(*zentaiteki keikaku keizai*). In the latter, the state manages every aspect of the economy, whereas in the former, the state manages only the important sectors.

As officials explained, the key difference between the partial planned economy and controlled economy was that in the "controlled economy," reforms are implemented within the accepted framework of the capitalist system. In the partially planned economy, the remnants of a liberal economy may be tolerated temporarily, but eventually the economy evolves toward a totally planned system.[57] Although Manchuria's planners adopted the basic techniques of the planned economy in terms of their extensive reliance on Soviet-style five year plans, planning directives, and centralized state control of the economy, their policies differed in at least one important respect. The Soviet economy was based on the socialist ideal of a classless society. In Manchukuo, leaders simply aimed to exclude monopoly capitalism through state control of the important heavy industrial sectors. In other sectors, they permitted Japanese residents and influential Chinese elites to enjoy capitalist profits and privileges over the masses while claiming to construct a moral world and model kingly way state based on racial harmony.

Despite claims of Manchurian independence, its economy was conceived as part of the so-called Japan-Manchuria bloc. Reflecting the concerns of the right, the army sought to prevent "mainland" domination of Manchuria by adopting the principle of "suitable site for suitable industries" (*tekichi-tekigyōshugi*). First officially proclaimed in the Kwantung Army's "Outline of Japan-Manchuria Economic Control Policies" of March 1934, this slogan reflected the army's determination that Manchuria not be an appendage of Japan's capitalist system. Manchuria's planners rejected the principles on which they perceived Japan's liberal capitalist system to be based: profit orientation, zaibatsu domination, and conflicts between capitalists and laborers. They planned industrial development based on a "rational division" (*gōriteki bungyōka*) of industries between the two countries. Rational division meant not economic development in the form of "colonial exploitation" by using Manchuria as a supply base for agricultural goods and raw materials as in the past. It represented a new Japan-Manchuria bloc in which industry would be developed where local conditions were most advantageous to that particular industry. Moreover, rather than assessing the profits of each industry, the profit of all the various industries in both countries would be assessed together as one sum and industries would be distributed so as to increase the total sum.[58] Planners highlighted the political character of the

---

57. Ibid., 322.
58. Minami Manshū tetsudō kabushikigaisha keizai chōsakai, ed., *Manshū keizai tōseisaku an* vol. 2, *Seisakuhen* (June 1932), 76.

economy in their repeated assertions that the bloc was built not for the interests of capitalists, laborers, entrepreneurs, or salaried employees, but rather for the nation as a whole and for the purpose of building a national defense economy.[59] Later, these principles were applied in planning for the Greater East Asia Co-Prosperity Sphere under Konoe's New Order movement.

The policy of "suitable site for suitable industries" existed more in theory than in practice in the early period. One scholar described the Japan-Manchuria bloc as less a truly fused body than an "economic union" or "linked economy" (*ketsugō keizai*). Wary of mainland intrusion into the prerogatives of the army, Kwantung Army leaders were keen on first establishing Manchuria on an equal and separate footing. The outline reflected this tentative approach toward creating a Japan-Manchuria bloc. Although it took "the rationalization and fusion of the East Asian economy" as its long-term goal, for the time being, it described the present economic relationship between Manchuria and Japan in terms of "interdependence" (*sōgo-ison*) and encouraged a relationship of "mutual aid" (*sōgo-fujo*) to be formed. While certain resources such as coal, iron, wool, and cotton were to be developed in Manchuria, there was as yet no comprehensive plan to achieve self-sufficiency involving the creation of synthetic substitutes for the resource deficiencies in both countries.[60]

In order to prevent the domination of monopoly capitalists of important industries, the state established the so-called "special company system." Under this system, the state took control over industries designated as "vital for national defense" or "in the public interest." Based on the principle of "one industry, one company" (*ichigyō issha*), the state created one company per industry. The concept of the special companies itself was not new—Mantetsu being itself a special company. What was new was the systematic use of special companies as the basic enterprise system. The Army created the first special companies in areas that provided basic services to the new state or provided the infrastructure for industrial development, such as the Manchurian Central Bank (1932) and the Manchurian Telegraph and Telephone Company (1933). The bulk of the special companies were established between 1934 and 1936.

The special companies reflected the army's technocratic view that an economy mobilized for total war was incompatible with an economy based on the pursuit of profits. Companies geared toward total war should possess a state character (*kokusakusei*), be large and comprehensive (*daikibo sōgosei*), and be rapidly developed (*kyūhakusei*). For this reason, the traditional form of a private stock corporation was inappropriate because "private enterprise is a form of free

---

59. Ibid.
60. Fujiwara, *Manshūkoku tōsei keizai ron*, 97–99.

economic enterprise essentially based on the pursuit of profits....The choice of industry, scale, and location (*ritchi*) and the pace of development are completely determined by the profit principle."[61] In other words, what was profitable for business was not necessarily good for the state. In contrast, Manchurian construction was described as possessing a "national planning character" in "going beyond the pursuit of profits and seeking to attain state goals."[62] The goal was not profits, but a rational determination of the industries that need to be developed, and the appropriate scale, location, and pace of construction of these industries based on Japan's national defense goals.

At the same time, Manchuria's planners were keen on avoiding direct state control. That the special companies took the form of stock corporations rather than state enterprises attests to the army's rejection of a communist or socialist-type system. In their early plans, Miyazaki and his colleagues at the Economic Research Association strived to strike a balance between what they termed the "direct control of management" (*chokusetsu keieiteki tōsei*) and "regulatory control of management" (*hōseiteki tōsei*). As they explained, in the case of purely state-owned enterprises "every aspect down to the last detail is directly controlled by the state." The result was that managers of such enterprises become essentially bureaucrats. Most have insufficient experience in business and tend to adhere to rules, become inefficient, and lack the incentive to produce profits. In contrast, in a system based on regulatory control, control was implemented only in the broadest sense. The state established the boundaries within which the managers can freely operate. It expanded the areas of activity in which they can fully exercise their ability. However, because the degree of control was weak, there existed the danger that control becomes insufficient. As a compromise between the two types of control, planners preferred "administrative control" (*kanriteki tōsei*), which they described in the terms of "supervisory," "guiding," or "indirect control" (*kantokuteki, shidōteki, kansetsukeieiteki tōsei*) of management.[63]

Civilian Technocrats and Military Fascism

Shifting our attention to Japan, how did civilian technocrats respond to military fascism? For these professionals, crisis meant opportunity—the chance to demonstrate their technical expertise and apply rational, problem-solving methods to the urgent problems at hand. Civilian technocrats took the high ground,

---

61. Manshūkoku seifu, *Manshū kenkoku jūnenshi*, 554.
62. Ibid.
63. Minami Manshū tetsudō kabushikigaisha keizai chōsakai, *Manshū keizai tōsei hōsaku*, 70.

condemning both the self-serving activities of capitalists and party politicians and the rash and violent tactics of the military and the radical right. They sought to steer the direction of national policy by promoting their own, rational conceptions of state reform. In contrast to the young officers' one-dimensional strategy of terror and intimidation, technocrats offered a multidimensional approach that addressed the various aspects of the crisis and their interrelationships within the given, evolving whole.[64] These groups offered solutions from the various professional standpoints of bureaucrat, business manager, engineer, and social scientist.

Leveraging on the military's blueprints for a Shōwa Restoration and broad-based national defense state, reform bureaucrats advanced their own managerial visions of efficient state coordination and economic control. They attempted to address what they perceived to be the main defects of liberal capitalism: excess competition, waste and inefficiencies in the production and distribution process, and labor conflict—put simply, economic disorder and insufficient government control over the economy. At the heart of Japan's economic problems, they believed, was the focus on individual profit at the expense of public welfare. Commerce bureaucrats called for a fundamental reorientation of economic activity from the standpoint of the national economy. After creating the Important Industries Control Law, Kishi Nobusuke and his colleagues at the Commerce Ministry drafted a series of control laws for heavy industry, beginning with the Petroleum Industry Law of 1934 and Automobile Manufacturing Law of 1936. Communications bureaucrats advocated a more activist role for the state in the public services sector. Working with Suzuki Teiichi at the Cabinet Research Bureau, Okumura Kiwao drafted the first state control law for the electric power industry in 1935.[65] Their policies for a controlled economy were geared toward the "public interest," represented the first of a series of control policies for every industry, and introduced the idea of "separating management from capital."

Engineers eagerly responded to the military's calls for societal reconstruction and continental expansion by carving out a new role for technology and technology experts in national policymaking. Since the 1920s, a group of young engineers led by Miyamoto Takenosuke had sought to promote technology and raise the low status of technically trained civil servants (*gikan*) through their engineering association, the Japan Artisan Club (Nihon kōjin kurabu). They later expanded the association into a national policymaking organization, the Japan

64. On multidimensional thought in planning, see Karl Mannheim, *Man and Society in an Age of Reconstruction* (New York: Harcourt, Brace, 1954), 152–154.

65. Okumura Kiwao, "Denryoku kokuei no mokuhyō to gainen," in Okumura Kiwao, *Denryoku kokuei* (Tokyo: Kokusaku kenkyūkai, 1936), 16.

Technology Association (Nihon gijutsu kyōkai). Following the Manchurian invasion, these "technology bureaucrats" began to look to Manchuria for solutions to Japan's domestic problems as well as their own professional quandaries. As Miyamoto noted, "the way out [of the crisis] for Japan lay in Manchurian-Mongolian development."[66] In addition, Manchurian development would increase the demand for government engineers, whose departments had faced drastic cuts. As part of his retrenchment policy, Prime Minister Hamaguchi had proposed to reduce the number of civil engineers at the Home Ministry by 90 percent. For this reason, these bureaucrats saw Manchuria and Mongolia as their promised land and proclaimed: "The new sphere of Mongolia-Manchuria, which is reported to have abundant resources and an expanse of over 90,000 *ri* is a *mecca* for us, a splendid region that will save the technology community."[67] For technology bureaucrats at the Ministry of Communications such as Matsumae Shigeyoshi and Kajii Takeshi, Japan's continental expansion presented the perfect opportunity to develop telecommunications technology. From 1930 they began to develop telephone cable technology to create a strategic telecommunications line between Japan and Manchukuo.[68]

Manchurian development also provided attractive opportunities for the new zaibatsu. At a time when the monopoly capitalism of the old zaibatsu was under attack, the new zaibatsu's national orientation and expertise in munitions-related industries made them attractive partners to the military. With the backing of General Ugaki, then governor-general of Korea, the Nitchitsu zaibatsu expanded and diversified its businesses in Korea. Nitchitsu won out over Mitsubishi in bidding for the contract to develop the Changjin River. One justification used by Japanese leaders for choosing Nitchitsu was that it was not a traditional zaibatsu. Ōkōchi pursued his dream of science-based industry in Manchuria by establishing Manchukuo's Continental Science Board to advance scientific and technical research. He also served as an advisor to a number of technocratic planning agencies and consultative committees in wartime Japan such as the Shōwa Research Association, Cabinet Research Bureau, Commerce and Industry Deliberation Council, Temporary Industrial Rationality Bureau, and Science Deliberation Committee.

The new zaibatsu that most benefitted from the liberal crisis was Nissan. Its mining business profited from Finance Minister Takahashi's decision to go off

---

66. Ōyodo Shōichi, *Miyamoto Takenosuke to kagaku gijutsu gyōsei* (Tokyo: Tōkai daigaku shuppansha, 1989), 176.

67. Ibid., 177.

68. Daqing Yang, "An Empire of Technology: Engineers as Visionaries and Political Actors," unpublished paper, 1999.

gold. During the stock market boom in the early 1930s, the company was able to raise significant amounts of capital. By the late 1930s, Nissan was one of the largest new zaibatsu with interests spanning mining, automobiles, shipping, chemicals, electric power, rubber, marine products, and insurance. Ayukawa Yoshisuke was invited by Ishiwara Kanji and other army leaders to develop Manchurian heavy industry. In 1937, Ayukawa transferred his entire Nissan operations to Manchuria and became the president of the new Manchurian Heavy Industries Corporation and one of the most powerful figures in Manchukuo.

Finally, a group of social scientists, alarmed by the impulsive and violent tactics of radical military officers, readily offered constructive policy recommendations based on sound research and a grasp of worldwide trends. Through their participation in technocratic think tanks, they exercised a degree of influence on politics not normally available to them through their academic posts. One such organization was the National Policy Research Association (Kokusaku kenkyūkai) founded by Yatsugi Kazuo in 1933. Yatsugi was a former member of the Cooperation and Harmony Society (Kyōchōkai), a semi-governmental organization created in 1919 to mediate between labor and management. Through the new association, Yatsugi sought to bring together the nation's leading scholars, bureaucrats, and military officers to draft comprehensive national policy in a wide range of areas.[69] Through his close ties with Ikeda Sumihisa at the Military Affairs Bureau, his organization directly participated in the drafting of the 1934 army pamphlet.

In the same year Gotō Ryūnosuke, with the backing of Prince Konoe Fumimaro, established the Shōwa Research Association (Shōwa kenkyūkai). The association brought together elite scholars, bureaucrats, business leaders, party leaders, and political personalities. Rōyama Masamichi directed its research program in the early years. Reflecting his training in administrative science and interest in the theory of functionalism of G.D.H. Cole, he looked to a strong and unified state as the primary agent of political and economic reform. He called for the creation of an "economic general staff" to overcome party politics and bureaucratic sectionalism—a view then advocated by Matsui Haruo, Resources Bureau chief and future member of the Shōwa Research Association. Although the research association did not issue concrete policy recommendations in the early years, its general stance was to reject both the current party system and the illegal and violent tactics of the radical right and seek statist solutions to the economic and political impasse.

---

69. Yatsugi Kazuo interview, in Nakamura Takafusa, Itō Takashi, and Hara Akira, eds., *Gendaishi o tsukuru hitobito*, vol. 4 (Tokyo: Mainichi shinbunsha, 1973), 88.

In the first half of the 1930s, civilian reformists were unpopular. The new technocratic organizations such as the Resources Bureau, Cabinet Deliberation Council, and Cabinet Research Bureau and their control policies were fiercely resisted by business and party leaders. Reformist proposals for an Electric Power Control Law caused such uproar among utility company owners that its principal author, Okumura, was sent abroad until tempers cooled.[70] In the wake of the February incident, control bureaucrats were viewed with increasing suspicion and fell out of favor with senior bureaucrats. The incoming commerce minister, Ogawa Gōtarō, was a member of the Minseitō and represented Osaka business interests. One of his first acts was to check reformist power by transferring Kishi to Manchuria and his boss Yoshino to northern Japan.

The military's fascist movement for reform had a profound impact on Japanese wartime politics but in less obvious, straight-forward ways. Certainly the violent actions of radical young officers succeeded in terrorizing the ruling establishment and abruptly bringing an end to party government. But in the first half of the 1930s, military and civilian technocrats lacked the political authority, institutional support, and unified vision to assume the reins of power in Japan and effectively challenge the status quo. Their plans to reform Japan's state and economy were fiercely attacked and sabotaged by business and party leaders. More than their initial attempts at internal reform, external reform via the Manchuria venture facilitated the rise to power of technocrats. The army's occupation of Manchuria and escalation of conflicts in North China gradually shifted Japan to a wartime footing by the late 1930s and increased the stature and influence of technocrats. War mobilization and empire building created new projects, planning agencies, and think tanks for civilian technocrats, especially reform-minded bureaucrats, industrialists, engineers, and social scientists. Moreover, the initial exclusion of the zaibatsu from Manchurian development enabled the army and its middle-class supporters to experiment with radical, anticapitalist policies and build their own political networks and planning organs.

The Manchurian venture also provided an opportunity for technocrats to develop a new political vision. Manchuria served as a social laboratory for Japanese fascism. In the early 1930s, Kwantung Army leaders produced their own vision of military fascism. In the second stage of Manchurian development from 1933, incoming Japanese bureaucrats gradually developed their own bureaucratic vision of techno-fascism. It is to this vision that we now turn.

---

70. Okumura, "Denryoku mondai kaiketsu no kagi," Interview in Andō Yoshio, ed. *Shōwashi no shōgen,* vol. 3 (Tokyo: Hara shoten, 1993), 161.

3

BUREAUCRATIC VISIONS OF
MANCHUKUO, 1933–39

With the arrival of elite Japanese bureaucrats in Manchuria from 1932, Manchurian development entered a new phase. These bureaucrats introduced a managerial dimension to the Kwantung Army's experiment in state reform. They devised new concepts, techniques, and institutions of planning and control that reflected the latest technocratic trends in interwar Japan, Germany, Soviet Russia, and the United States. During the first year, the Kwantung Army constructed the basic framework for fascism: a military-dominated, totalist state, pan-Asianist vanguard party, and planned economy. Building on this framework, bureaucrats under the leadership of Kishi Nobusuke refashioned the various organizations of the state and linked them more tightly together into an integrated system of technocratic coordination and control. These bureaucrats modified the army's Soviet-style controls and incorporated a more business-oriented, managerial approach toward economic planning. The ideology infusing this system was no longer the "Dairen ideology" of Mantetsu and the army's socialistic vision of a "broad-based national defense state" but the "Shinkyō ideology" of bureaucrats and their techno-fascist vision of a high-performance, "advanced national defense state."[1] As one Mantetsu researcher bitterly noted, the ideology of the new capital was no longer "ethnic harmony" but "the rising sun is your boss" (*ora ga oyakata wa hi no maru da*).[2]

---

1. Kusayanagi, *Jitsuroku Mantetsu chōsabu*, 248.
2. Ibid., 251.

Laying the Foundations of Planning

Setting the Terms

Bureaucratic participation in the Manchurian venture began with the change in Kwantung Army leadership. Within one year of the Manchurian invasion and founding of the new state, senior Kwantung Army leaders Honjō Shigeru and Ishiwara Kanji were called back to Japan. Senior officers in Tokyo were concerned about the alienating effects of the Kwantung Army's widely publicized anti-zaibatsu stance and its independent behavior. Japan's Army Ministry sought to regain control of Manchuria policy by appointing senior officers General Mutō Nobuyoshi as Kwantung Army commander in chief and Lieutenant Colonel Koiso as Kwantung Army chief of staff in August 1932. There were clear differences between the old and new leadership. In contrast to Ishiwara's faction, which embraced the antiestablishment, pan-Asian ideals of the radical right, the incoming army leaders more closely followed the policies of mainland interests. Mutō and Koiso adopted a pragmatic, low-key approach. They walked a fine line between encouraging zaibatsu investment on the one hand, and quietly preparing the way for a planned economy on the other. The higher rank of the new Kwantung Army leaders reflected a rise in authority and standing of the Kwantung Army within the Japanese military and government. Under Mutō and Koiso, a new "three-in-one" system was adopted by the army in which the Kwantung Army commander in chief also held the posts of ambassador extraordinary and plenipotentiary and governor of Kwantung Province.[3]

Faced with the task of economic development, the new Kwantung Army leaders began to recruit mainland bureaucrats to Manchuria. Among the first to arrive were bureaucrats from the Ministry of Finance. In July 1932, Hoshino Naoki, a Finance Ministry veteran, was appointed as advisor to Manchukuo's Finance Department (zaiseibu) and head of that department's secretariat. Hoshino then recruited junior finance bureaucrats to Manchuria to manage the important areas of accounts, finance, and customs. His strategy was to choose promising young bureaucrats who could assume the equivalent posts in Japan ten years later.[4] Among the bureaucrats he selected were Furumi Tadayuki, Matsuda Reisuke, and Mōri Hideoto. Furumi and Matsuda were placed within the Accounts Section (shukeisho) of the General Affairs Agency in 1932, with Matsuda as head of Accounts, and Furumi as head of that section's secretariat and also head of the

---

3. For an examination of this policy, see Y. Tak Matsusaka, "Managing Occupied Manchuria," Peter Duus, Ramon H. Myers, and Mark R. Peattie, eds., *Japan's Wartime Empire* (Princeton: Princeton University Press, 1996), 112–120.

4. Hoshino Naoki, *Mihatenu yume* (Tokyo: Daiyamondosha, 1963), 9.

Special Accounts Section. Hoshino invited Mōri to come to Manchuria a year later, on the request of Hoshino's university classmate Kamei Kan'ichirō.[5] Mōri succeeded Furumi as head of the Special Accounts Section. All three served in key posts within the Manchukuo government. As Hoshino's right-hand man, Furumi served in various capacities at the General Affairs Agency and later became a director of the Concordia Association and deputy chief of the Economic Department and General Affairs Agency. Matsuda headed the important Planning Committee and later succeeded Furumi as deputy head of the Economic Department. In 1941 he returned to Tokyo to work in the Cabinet Planning Board. Mōri, who was in charge of national tax policy for Manchukuo, became an economic advisor in North China in 1938. In 1939, he returned to Japan and joined the new Asia Development Board (Kōain). After a financial system had been created, bureaucrats were recruited for the task of industrial development. The Kwantung Army had requested Kishi, a rising star at the Ministry of Commerce and Industry. However, his boss Yoshino was reluctant to let him go and another commerce bureaucrat Takahashi Kōjun was sent in his place. Kishi finally did transfer to Manchuria in October 1936. In the meantime, among the young bureaucrats he arranged to be sent were Shiina Etsusaburō and Minobe Yōji. In 1933, Shiina was appointed chief of the planning section in Manchukuo's Industrial Department and Minobe as secretary of the Patent Bureau and head of inspection.

The first bureaucrats to arrive in Manchuria viewed their new assignments with considerable trepidation. Many were concerned about the violence in the region, harsh climate, working conditions, and salary, which had not been established before their arrival.[6] Moreover, they were uneasy about the prospects of collaboration with the Kwantung Army, which had the reputation in Japan of being tyrannical and defiant of authority (gekokujō). On his way to meeting Kwantung Army leaders for the first time, Hoshino "imagined how fearful and agitated these army men would be at a time when fighting was going on in every region of Manchuria."[7] Furumi dreaded working with the notorious Manchukuo police chief Amakasu Masahiko, who had brutally murdered the noted anarchist Ōsugi Sakae in a Japanese prison. At first Furumi, whose family had close ties with the Amakasu family, had resolved not to meet Amakasu using the letter

5. Kamei Kan'ichirō, "Kamei Kan'ichirō danwa sokki roku" (Tokyo: Nihon kindai shiryō kenkyūkai, 1969), 37.

6. Hoshino, Mihatenu yume, 14; Shiina Etsusaburō tsuitōroku kankōkai, ed., Kiroku: Shiina Etsusaburō, 2 vols. (Tokyo: Shiina Etsusaburō tsuitōroku kankōkai, 1982), 115; Furumi, Wasureenu Manshūkoku, 32. Furumi claims that the government did not even provide travel costs to Manchuria in the beginning.

7. Hoshino, Mihatenu yume, 20–21.

of introduction which his father had prepared for him.[8] Some bureaucrats re-
luctantly accepted assignments in Manchuria after being subjected to pressure
tactics by their superiors and colleagues. Hoshino turned down requests to go to
Manchuria three times on account of family difficulties, and finally succumbed
after a colleague convinced Hoshino's wife to get him to accept.[9] Others, such as
Tanaka Yasushi, Nagai Tetsuzō, and Harada Matsuzō, readily took up the offer.

These bureaucrats arrived with a fundamentally different attitude toward
Manchurian development. Reflecting their view of politics as one of rational
administration and their materialist interpretation of national interests, these
bureaucrats understood kingly way as efficient bureaucratic administration and
the creation of industrial wealth. They saw Manchukuo as representing an administrative showcase to the West and an unprecedented administrative challenge. Hoshino argued that "it would be terrible if half-measures were taken and
mistakes made toward the other country [Manchukuo] early on. Japan would
be disgraced forever in the eyes of the world, and in the end this would cause
much trouble to our country."[10] Shiina, who at the time of the invasion had been
on a study-tour abroad, later recalled the skeptical attitude of the international
community about whether Japan would be able to control Manchuria through
means other than just military force.[11] While in Manchuria, Shiina took to heart
the advice he received from one Manchurian bureaucrat: "As for preserving peace
in Manchuria, there is a 7:3 ratio of military might to government. Without a 7:3
[reversed] ratio of government to military might it cannot be achieved. To the
extent possible, win the hearts of the people through the power of government;
if this fails, then bring out military force."[12]

The esprit de corps and privileged status of the incoming bureaucrats clashed
with the official policy of ethnic harmony. Kwantung Army leaders had been
concerned to preserve the political ideals of kingly way and ethnic harmony.
They made the following request to the first group of bureaucrats: "We would
like you to be prepared to practice ethnic harmony—to adopt the attitudes of
the Manchurians, wear the same clothes as the Manchurians, and eat the same
food."[13] Already in its institutionalization of a dual system of formal Chinese and
actual Japanese rule, the Kwantung Army had greatly modified the meaning of
"ethnic harmony." They justified this arrangement on the grounds that the local

---

8. Furumi, *Wasureenu Manshūkoku*, 183–184.

9. Ibid., 29.

10. Ibid., 4.

11. Shiina Etsusaburō, *Watakushi no rirekisho*, vol. 41 (Tokyo: Nihon keizai shinbunsha,
1970), 186.

12. Ibid., 192.

13. Manshūkokushi hensan kankōkai, ed. *Manshūkokushi: Sōron*, 246–247.

Chinese bureaucrats trained in the Qing tradition were not qualified to run a modern state. For the time being, Japanese bureaucrats were placed within the various departments but their numbers were limited. Fixed ratios between Japanese and local bureaucrats were established in each department: 7:3 within the General Affairs Agency; 5:5 within the Business Department; 3:7 within the Department of Education.[14] With the arrival of mainland bureaucrats, however, the policy of ethnic harmony became increasingly difficult to uphold. Japanese bureaucrats received preferential treatment in terms of compensation, living conditions, and training, which generated considerable resentment among local bureaucrats.[15] Salaries for Japanese bureaucrats in Manchuria were more than double those in Japan; in contrast, salaries of Manchurian bureaucrats were miniscule.[16] Furumi, who was responsible for establishing the compensation scheme for all Manchukuo bureaucrats, claimed to have tried to lower the salaries of Japanese bureaucrats to bring them in line with local levels, but was only able to push through provisional reductions.[17] Japanese bureaucrats also obtained substantially better accommodations and meals. Finally, separate career tracks were instituted for Japanese and local bureaucrats via the new training institute for Manchukuo bureaucrats, the Great Unity Academy (Daidō gakuin). Established in 1932, the school aimed to train young Japanese and Manchurians for the regional offices and to promote racial harmony and a common consciousness among them. Despite its stated goals, however, the school soon introduced separate curriculums for Japanese and local students. By 1936, Japanese department and bureau heads such as Shiina were appointed to teach Japanese students, and the school became the official training ground for the Japanese bureaucratic elite in Manchuria.[18] The reform bureaucrat Kan Tarō later recalled how Japanese bureaucrats were indoctrinated at the Great Unity Academy in the ideas of ethnic harmony and kingly way being promoted by such groups as Kasagi's Great Majestic Peak Association and Ishiwara's East Asian League.[19]

---

14. The Department of Education was created out of the education section within the Department of Civilian Affairs in July 1932.

15. For a discussion of the treatment of Chinese bureaucrats in Manchukuo, see Hamaguchi Yūko, *Nihon tōchi to tō-Ajia shakai* (Tokyo: Keisō shobō, 1996).

16. Moreover, in the early years Japanese bureaucrats were paid in silver-based yuan and their salaries in yen soared after the United States drove up the price of silver.

17. Katakura and Furumi, *Zasetsushita risō koku: Manshūkoku kōbō no shinsō*, 205–206.

18. Manshūkokushi hensan kankōkai, ed. *Manshūkokushi: Sōron*, 251–255.

19. Kan Tarō, Interview, in Nakamura Takafusa, Itō Takashi, and Hara Akira, eds., *Gendaishi o tsukuru hitobito*, vol. 1 (Tokyo: Mainichi shinbunsha, 1971) 247.

Bureaucrats represented a new and formidable contender for power within the new state. Wary of bureaucratic encroachment into their prerogatives, the Kwantung Army adhered to its dual strategy of permitting bureaucrats to manage the state and establish the foundations of Manchurian industry while exerting indirect control through internal guidance. The Kwantung Army monitored the activities of bureaucrats through regular communication channels between the army and bureaucracy. The main vehicle for the army's strategy of indirect control was the newly reestablished Third Division within the Kwantung Army Staff, which was charged with overseeing bureaucratic activities. The Third Division (later the Fourth Division) served as the main contact point to the General Affairs Agency, which now had to obtain the army's formal approval on all important matters such as laws, personnel appointments, and agendas for the weekly meetings of the General Affairs Agency.[20] The army's Special Affairs Unit was also expanded and headed by Koiso Kuniaki, a recognized authority on total war mobilization.

The Kwantung Army's policy of internal guidance drew much criticism in Japan not only from bureaucrats, but also the political parties, ministries, and army headquarters. Nagata Tetsuzan reportedly warned Kwantung Army leaders to refrain from interfering in Manchukuo's political affairs, since it lacked administrative experience, and allow the state to develop independently.[21] Army minister Hayashi Senjūrō also publicly proclaimed his opposition to further Kwantung Army intervention in Manchukuo's administration and economy.[22] Kishi demanded that the army abstain from internal guidance over his activities. He recalled setting forth the conditions for his participation to Kwantung Army chief of staff Itagaki as follows:

> I did not come here for lack of work in Japan. The army meddles in industry-related matters without any understanding. It is also not good that due to the authority of the Kwantung Army, tradesmen and various people from industry are hanging around the corridors of Kwantung Army headquarters. As for myself, I will take instructions from the Kwantung Army with regard to Manchukuo's larger political principles, but I would like to have individual matters of industry and business entrusted to me....If my request cannot be granted, I will return because the reason for my coming to Manchuria no longer exists.[23]

20. Mutō Tomio, *Watakushi to Manshūkoku* (Tokyo: Bungei shunjū, 1988), 113.

21. Yatsugi Kazuo, *Shōwa dōran shishi* (Tokyo: Keizai ōraisha, 1978), 116; on the Kwantung Army's decision to leave administration to incoming bureaucrats, see Kusayanagi, *Jitsuroku Mantetsu chōsabu*, 250.

22. Yoshida, "Gunji shihai: Manshū jihen ki," 150.

23. Kishi, Speech at Bunyūkai, May 9, 1980, 8–9.

Bureaucratic Planning Apparatus

Bureaucrats eagerly took up the challenges of managing the new bureaucracy and laying the foundations of Manchurian industry. Soon after their arrival, they began to build their own bureaucratic planning apparatus. Shiina's first project was to create a new research bureau. He disparagingly remarked that the current planning section which he had just taken over was "a place for people with nothing to do…staffed by Manchurians who spent their days practicing their calligraphy on old newspapers."[24] As for Mantetsu's research capabilities, Shiina found that it had collected data on only those areas attached to the railroad in order to estimate railway revenues but not on areas related to Manchurian industrial development. Shiina established the Temporary Industrial Research Bureau (Rinji sangyō chōsakyoku) within the planning section and became its first director. The institute marshaled a staff of three hundred and fifty to four hundred members at its peak and was composed of bureaucrats from Japan's ministries of Commerce and Industry, Agriculture and Forestry, Government-General of Korea, Mantetsu researchers, and prominent specialists from Japan.[25] Research covered a vast area ranging from land and river surveys for the development of industry and timber resources to research on rice cultivation, hydroelectric power, mining, minerals, and transportation methods. In addition, the research bureau conducted studies comparing the legal framework for economic planning in Japan, Germany, and America.[26]

At the same time, Hoshino formulated plans for a state science institute, the Continental Science Board (Tairiku kagakuin). Hoshino commissioned Riken president Ōkōchi Masatoshi, a noted authority on the Japanese armaments industry and a consultant to Japan's Resources Bureau.[27] Established in March 1935, the Continental Science Board became the central research organ for science and technology. Based on the Soviet model of centralized research, it was designed to mobilize science and technology for the purposes of the state. Research projects were decided by a Science Deliberation Committee (Kagaku shingikai), composed of the prime minister, deputy head of the General Affairs Board, department and bureau heads, the Continental Science Board's director, and academic

---

24. Shiina Etsusaburō, Interview, in Nakamura Takafusa, Itō Takashi, and Hara Akira, eds., *Gendaishi o tsukuru hitobito*, vol. 4 (Tokyo: Mainichi shinbunsha, 1971), 264; Shiina, *Watakushi no rirekisho* 41, 188.

25. Shiina, *Gendaishi o tsukuru hitobito*, 265.

26. Rinji sangyō chōsakyoku, "Kakkoku sangyō tōseihō no shiteki hattatsu" (October 1934), Minobe Yōji Documents.

27. Hoshino, *Mihatenu yume*, 172.

specialists.[28] The Continental Science Board became an important model for the Technology Board, which would serve as the central organization of the Science-Technology New Order movement in Japan in the early 1940s.

Both the Kwantung Army and Mantetsu perceived both research organs as a threat and feared that they would make their own planning activities redundant. In a meeting between Kwantung Army officers and researchers from Mantetsu's Economic Research Association in June 1935, both groups voiced their concerns about the new developments. On the topic of Shiina's research bureau, one Kwantung Army officer, Colonel Harada Kumakichi, remarked: "When Manchukuo's Industrial Research Bureau was created it became a problem within the army after all. We had thought that we could maintain control if we carried out internal guidance vis-à-vis the Industrial Research Bureau within the army, while keeping an eye on the Economic Research Association. In fact, however, isn't it so that this internal guidance hasn't worked well?"[29] In response, researchers of the Economic Research Association appealed to Kwantung Army officers to take control of the situation. The Continental Science Board duplicated the scientific activities of Mantetsu. One researcher remarked with apparent exasperation: "Within Mantetsu, there is a geological survey bureau. There are about forty technical specialists who conduct general geological surveys. On the Manchukuo side, the Continental Science Board was created and intends to do the same thing. Moreover, the Industrial Research Bureau also plans to do the same thing."[30] Their concerns were not groundless. In the case of Mantetsu's geological survey bureau, all its research findings, facilities, and staff in Dairen were soon transferred to the new Continental Science Board located in the capital city of Shinkyō.[31] Through the new research bureau and science institute, the reform bureaucrats assumed control over important aspects of industrial planning.

Bureaucrats established a new planning office to serve as the brain trust of the General Affairs Agency director and as the central organ of the new planning apparatus.[32] According to Furumi, he and Mōri had originally envisioned a Planning Bureau (Kikakukyoku) similar to Japan's Cabinet Investigation Bureau to deliberate and decide on national policy. As Furumi explained, "In Manchukuo the General Affairs Agency's Accounts section had determined policies regarding

28. Janis Mimura, "Technocratic Visions of Empire: Technology Bureaucrats and the 'New Order for Science-Technology,'" in Harald Feuss, ed., *The Japanese Empire in East Asia and its Postwar Legacy* (Munich: Iudicium Verlag, 1998), 107–110.

29. Minami Manshū tetsudō kabushikigaisha keizai chōsakai, "Daiikkai kantōgun bakuryō, keichō dankai kiroku" (June 1935) (gokuhitsu), 11.

30. Ibid., 12.

31. Hoshino, *Mihatenu yume*, 174.

32. Yamamuro, "'Manshūkoku' tōchi kateiron," in Yamamoto Yūzō, ed., *"Manshūkoku" no kenkyū* (Tokyo: Ryokuin shobō, 1995), 107.

disbursement of funds and handling of financial measures. However, in terms of actual deliberation and research on various policies, there was no organ to handle this."[33] Plans for the new bureau were actively supported by Akinaga Tsukizō, a recent appointee to the Kwantung Army's Third Division who became a close associate of the reform bureaucrats. Due to opposition from other government planning units which opposed reduction of their own area of authority, the establishment of the bureau was delayed until November 1935 and its authority curtailed. The new organization was placed within the General Affairs Agency as a smaller Planning Section (Kikakusho). Bureaucrats later strengthened this unit by creating a Planning Committee (Kikaku iinkai) in 1938 and a full-scale Planning Bureau in 1945.

With the establishment of the Manchurian Affairs Bureau in 1935, the Kwantung Army now had to share power with bureaucrats with regard to coordinating policy between Manchukuo and Japan.[34] Some historians argue that the creation of the Manchurian Affairs Bureau represented a strengthening of Kwantung Army power. The bureau, which was headed by the army, enabled it to consolidate its authority over the foreign and colonial ministries in Manchurian affairs.[35] When one compares the policymaking route before and after the establishment of the new bureau, however, it becomes evident that new players were being incorporated into a process previously controlled by the army. In the past, the Kwantung Army had communicated with the Japanese government via the Army Ministry and transmitted inquiries to the army's Military Affairs Bureau. The head of this bureau then forwarded the matter to the army minister, who contacted the relevant ministries. After the establishment of the Manchurian Affairs Bureau, however, this policymaking route became less direct. Now the Army Ministry, having received an inquiry from the Kwantung Army, forwarded the matter to the Manchurian Affairs Bureau, which then contacted the various ministries. Once an understanding had been reached between the bureau and relevant ministries, the decision was forwarded by the bureau back to the army, the Kwantung Army, and Manchukuo government.[36] The Manchurian Affairs Bureau was composed of military and civilian technocrats from the various ministries and agencies in Japan. Among its members were powerful reform bureaucrats such as Kishi, Yoshino, Okumura, Aoki Kazuo, and Matsui Haruo. In the 1940s, at the height of the New Order movement, other prominent reform bureaucrats

33. Furumi, *Wasureenu Manshūkoku*, 101–102.
34. Yoshida, "Gunji shihai: Manshū jihen ki," 153–154; Yamamuro, "'Manshūkoku' tōchi kateiron," 108–109.
35. Matsusaka, "Managing Occupied Manchuria," 127–135.
36. Katakura and Furumi, *Zasetsushita risō koku*, 210.

joined the Manchurian Affairs Bureau. One can surmise that through the place-ment of such figures within key posts on both sides of the policymaking route in Japan and Manchuria, the reform bureaucrats were able to wield considerable influence in Manchurian affairs.

In addition, as the Manchurian counterpart to the new bureau, the Japan-Manchuria Economic Joint Committee (Nichi-Man keizai kyōdō iinkai) was established in Shinkyō in June 1935. The committee handled all important economic matters arising between Japan and Manchuria including supervising Japanese-Manchurian joint management of the special companies in Manchu-ria.[37] According to Kwantung Army officer Katakura Tadashi, economic problems were handled directly between advisors representing each side rather than via the Kwantung Army's Special Affairs Unit, which up until now had played the central role.[38] Shortly afterward, the Special Affairs Unit was dissolved.

The group that was most severely disadvantaged by the new political recon-figuration was Mantetsu, which saw its research role gradually diminished in the planning process. Members of its Economic Research Association staff no longer enjoyed free access to government data and were channeled into "peripheral" research assignments. From the second half of 1934, they began to shift their research focus from Manchuria to China and the Soviet Union and to less politi-cally sensitive projects such as topographic studies of military-controlled areas. The Economic Research Association increasingly served as no more than an agent of the army's internal guidance, assuming the task of double-checking and assessing the policy drafts of bureaucrats for the army. According to a research staff member Miwa Takeshi, the army's policy of internal guidance had caused considerable discontentment among "proud" bureaucrats and a deterioration of relations between certain bureaucratic departments and the Economic Research Association.[39] Miwa described his experience of trying to work with the reform bureaucrats as follows:

> In 1936, I explained the research on the price structure of important export and import commodities and the commodity distribution sys-tem to Shiseki Ihei, of the Ministry of Commerce and Industry and [now] head of the commerce unit at the Temporary Industrial Research Bureau of Manchukuo's Business Department. Mr. Shiseki rejected my requests for cooperation, saying that they would conduct the industry

---

37. Manshūkokushi hensan kankōkai, ed. *Manshūkokushi: Sōron*, 388.
38. Katakura Tadashi, "Manshūkoku keizai seisaku no genjitsu to shōrai ni tsuite" (Tokyo: Nihon jitsugyō kyōkai [December 1935]) 10.
39. Imura Tetsuo, ed., *Mantetsu chōsabu: kankei sha no shōgen* (Tokyo: Ajia keizai kenkyūjo, 1996), 376, 451.

research needed by the Manchukuo government and didn't need the assistance of the Economic Research Association. I was taken aback by this bureaucratic insolence. Eventually, research on the price structure of important export and import commodities were carried out jointly with Manchukuo's Finance Department. In addition, on the request of the Kwantung Army General Staff's Third Division, the Economic Research Association drafted the basic policy plan for Manchukuo's controlled economy. I received an earful about this from Shiina Etsusaburō, who headed the Control Section within the Business Department's secretariat. He declined my offers of data and collaboration in research.[40]

Bureaucratic Mass Mobilization

Incoming bureaucrats modified the army's policies for mass mobilization. The Concordia Association gradually assumed two functions within Manchukuo. It became the state's main propaganda arm and provided the basic infrastructure for regional government. Originally conceived as the elite vanguard party of the Kwantung Army, it evolved into a bureaucratic organ for regional control and mass mobilization. In its brief existence, the Concordia Association evolved from a "party-type political group" controlled by Ishiwara's national founding faction, to a bureaucratic "leadership-type organization" through which bureaucrats led the multiethnic state, and finally to an "assistance-type" organization that unified the government and people into one body.[41] Its official platform changed from adopting an independent, anticapitalist stance and promoting ethnic harmony and kingly way to espousing vague, universal expressions of the unity between the government and the people and of the "inseparable relationship between Japan and Manchukuo" (*Nichi-Man fukabun kankei*).[42] In its latter form, the Concordia Association served as a model for Japan's Imperial Rule Assistance Association under Konoe's movement for a Political New Order.[43]

Propaganda played a special role in the army's national defense state. In order to obtain the cooperation of the masses, army leaders sought to exert control not only through military force, but also by spiritual mobilization of the people using technologically advanced propaganda. In the early years leaders had

---

40. Ibid., 466.
41. Koyama, *Manshū kyōwakai no hattatsu*, 42.
42. Ibid.
43. Ibid. See also Mitani Taiichirō, "Manshūkoku kokka taisei to Nihon no kokunai seiji," in *Kindai Nihon to shokuminchi*, vol. 2, *Teikoku tōchi no kōzō*, vol. 2 (Tokyo: Iwanami kōza, 1992).

focused on obtaining the support of local Chinese elites and suppressing opposition to Japanese rule under its policy of "peace preservation." Now they turned their attention toward popular mobilization for total war. The challenge was to both gain popular acceptance of Japanese rule and mobilize their support for rapid industrialization.[44]

In 1932 the army had established the Manchukuo Wire Service (Manshūkoku tsūshinsha) to take over news gathering and broadcasting within Manchuria. In the fall of 1933 Okumura Kiwao and other communications bureaucrats, such as Tamura Kenjirō and Kajii Takeshi, established the Manchurian Telegraph and Telephone Company. This special company was responsible for overseeing communications facilities such as radio in Manchuria and coordinating communications services between Japan and Manchukuo. Through its control of radio broadcasting it became the state's main instrument for mobilizing support abroad and in Manchukuo.

In addition, bureaucrats established the administrative infrastructure for state propaganda activities. In April 1933, they established the Information Section (Jōhōsho) within Manchukuo's General Affairs Agency. This executive-level organization was designed to coordinate and manage the state's various propaganda agencies, which had taken over the propaganda functions of the recently dismantled National Affairs Bureau. It gradually expanded its powers to include state control over the press, film, and radio. In 1936, bureaucrats actively consolidated and increased state control over newspapers under a new administrative unit, the Manchurian Public Information Society (Manshū kōhōkai). Shortly after the outbreak of the China War, they established the Manchurian Film Society (Manshū eiga kyōkai) to create propaganda films for the state.

In March 1934 the government appointed Pu Yi as emperor of Manchukuo. Leaders hoped that the new imperial system would broaden the base of support among local power holders in Manchuria and thereby stabilize the political situation. Moreover, by linking it to Japan's imperial system, Manchukuo policies could be identified with those of the Japanese state. The Japan-Manchuria bloc, through which the early Kwantung Army officers had been careful to assert Manchukuo's independence vis-à-vis Japan, was now altered to represent an "inseparable unified body" between Japan and Manchukuo (*ittai fukabun kankei*) under the slogan of "Japan-Manchukuo: one virtue, one heart" (*Nichi-Man*

---

44. Manshūkokushi hensan kankōkai, ed., *Manshūkokushi: Kakuron*, 61. These organs included the Propaganda Section within the Department of Foreign Affairs, the Pacification Subcommittee within the Peace Preservation Committee, and the propaganda wing of the newly created Department of Education.

ittoku isshin). From 1935 onward, Manchukuo essentially became a part of "continental Japan" (*tairiku Nihon*).[45]

The fate of the Concordia Association hinged on the whim of the particular army leader. Originally, Ishiwara and his national founding faction had envisioned a leading role for the Concordia Association as the vanguard party of the Kwantung Army to guide state policy. Ishiwara's successor Koiso, however, disapproved of its strong, independent "political party" character. In order to strengthen the state's control over the Concordia Association, Koiso replaced its founding members Yamaguchi Jūji and Ozawa Kaisaku with Japanese bureaucrats under General Affairs Agency deputy chief Sakatani Kiichi. At this time, Manchurian leaders even considered dissolving the Concordia Association. Following the assumption of Minami Jirō to commander in chief, however, interest in the Concordia Association revived. Both Minami and his successor Ueda Kenkichi were sympathetic to the early ideals of Ishiwara's founding faction and viewed the Concordia Association as a useful vehicle for political mobilization. Furthermore, the arrival in Manchuria of two close associates of Ishiwara also helped mobilize support for the Concordia Association within the army.[46] Ueda laid out a new platform for the Concordia Association in an internal policy statement, "Concerning the Basic Character of the Concordia Association and the Fundamental Principles of Manchukuo." The document embraced the basic principles of Manchukuo's founders and described kingly way rule as government by sages, namely Kwantung Army leaders, and by a sagelike organizational body that reflected the people's will. Ueda described the Concordia Association as the "spiritual womb of the government" that united bureaucrats and the people.[47] After the next rotation in February 1937, however, the Concordia Association again lost its privileged status within the Kwantung Army. The new chief of staff Tōjō Hideki did not share the same enthusiasm of his predecessors. Even Ishiwara's return to Manchuria as deputy chief of staff of the Kwantung Army in September 1937 failed to revive the Concordia Association's original mission. Tōjō's pragmatic, methodical approach clashed with the brilliant, headstrong Ishiwara, who gradually lost out to Tōjō's faction.

Despite the abrupt shifts in the Kwantung Army's policies toward the Concordia Association, two trends remained constant: the increasing control of bureaucrats over its activities and its gradual transformation into an organ for regional state control. Through two major reorganizations, bureaucrats laid the grounds for the creation of a mass mobilization organization. The first reorganization in September 1934 aimed primarily at unifying and strengthening leadership at the

---

45. Fujiwara, *Manshūkoku tōsei keizai ron,* 61.
46. Ibid., 143.
47. Mitani, "Manshūkoku kokka taisei to Nihon no kokunai seiji," 198–199.

center. The Japanese government secured its control over the Concordia Association by replacing its leaders with bureaucrats. It created an institutional network to link the center and regions through the Union Council (Rengō kyōgikai). Established at the various levels of local government, the councils served as the main channel through which the regional branches conveyed their views to central leadership. Institutionally, they symbolized the "union of bureaucrats and the people" (*kanmin ittai*). At the Union Council, regional branch delegates were organized by occupation and ethnicity and convened with Japanese leaders, including Concordia Association leaders and advisors, government officials, and special company officials. The Union Council was designed as a forum in which branch representatives, the government, and participants of other related organs could exchange their views unreservedly and realize the ideal of "relating the imperial virtue" (*gotoku o noberu koto*) and "advancing the people's feelings" (*tamigusa no jinjō o jyōtatsu suru/min no jinjō o chōtatsu suru*). Sakatani described the Union Council's methods as "neither the eighteenth-century despotism that produced tyrants, nor the parliamentary government that brings the world to an impasse, nor even the fashionable fascist governments of recent days, but so-called kingly rule government."[48] Sakatani hinted that the Union Council would eventually become Manchukuo's legislative organ.[49]

The ideal and reality of mass politics in Manchuria lay far apart. Despite the institutional trappings and rhetoric of popular, local representation, the Union Council was essentially a bureaucratic organ controlled by Japanese officials. Concordia Association leaders rejected outright a popular system that acknowledged conflicting interests between the people and the government. Branch representatives were not selected through popular election, but appointed from among a small group of officials within each branch. Bills produced by the branches were first screened by powerful branch leaders within a committee to ensure that they conformed to state goals. The bills presented at the Union Council were not passed on the basis of majority vote. They were passed via a process of so-called "unanimous unification and sanction" (*shūgi tōsai*), by which the council chief would hear the opinions of the branch representative and government and related organs and then make a decision based upon a consideration of how best to unify the will of the state and people. According to kingly way propaganda, this council's decision would then be supported automatically by the people who would understand and trust the council chief.

The Concordia Association did not adopt mass mobilization as its main goal and actively extend its propaganda activities into the regions until the second reorganization in July 1936. In early 1937 its leadership structure was reorganized

---

48. Koyama Sadatomo, *Manshūkoku to kyōwakai* (Tokyo: Manshū hyōronsha, 1935), 530–531.
49. Ibid., 522.

and new officials were brought in. Amakasu and Furumi took over the leadership of the newly created General Affairs Section (Sōmubu) and Supervisory Section (Shidōbu). Both actively sought to build a mass organization in various ways. First they tried to eliminate once and for all Ishiwara's faction within the Concordia Association, which had hoped to form a vanguard party based on the ideal of "one party, one state." They revised the three-tiered membership structure and opened up membership to all residents in Manchuria over twenty years of age regardless of citizenship. Second, they created youth training organizations in order to extend the Concordia Association's regional presence and increase membership. The youth training centers (*Seinen kunrensho*) engaged in activities such as military drills, ideological indoctrination, and Japanese language workshops for youths between the ages of fifteen and twenty-five. To supplement these training centers, they created boys' and young men's groups for males between the ages of ten and twenty-five. By March 1939, there were 2,000 youth groups with 385,000 members, and 1,200 boys groups with 200,000 members.[50] In addition, Patriotic Voluntary Service Units (Giyuhōkōtai) composed of males between the ages of twenty and thirty-five were created in the cities, designated towns, and heavy industrial areas to serve as local self-defense and policing units. They reorganized the Concordia Association's branches on a regional basis rather than on the basis of ethnicity, occupation, or class as in the past. Whereas previously powerful landowners and merchants had been given considerable voice in regional administration and had formed the basis of the branches, now they were to be replaced with patriotic youth. Finally, they launched an active propaganda campaign. Concordia Association leaders sought to promote the state's ideals and unify the people using every medium available—from newspapers, magazines, pamphlets, and the Concordia Association's official journal, to lectures, radio, films, and plays.[51] In order to reinforce the party's mass character, Amakasu instituted the wearing of party uniforms for all members, regardless of rank or class.[52]

Bureaucratic Strategies of Control

From Industrial Rationalization to Industrial Control

Whereas the Kwantung Army and Mantetsu researchers had conceived of economic planning along the lines of the socialist command system, bureaucrats

---

50. Kazama Hideto, "Nōson gyōsei shihai," in Asada Kyōji and Kobayashi Hideo, eds., *Nihon teikokushugi no Manshū shihai: Jūgonen sensōki o chūshin ni* (Tokyo: Jichōsha, 1986), 284.

51. Manshūkokushi hensan kankōkai, ed. *Manshūkokushi: Kakuron*, 113.

52. Furumi, *Wasureenu Manshūkoku*, 150–152.

developed their approach toward economic control based on the concepts of industrial rationalization. From the time of the passage of the Important Industries Control Law in 1931 until his transfer to Manchuria in October 1936, Kishi actively promoted and elaborated his view of the national economy. During this period, he rapidly advanced within the ministry with the backing of his powerful mentor, Vice Minister Yoshino Shinji. Kishi's promotion from Industrial Policy section chief to head of Documents in December 1933 and to Industrial Affairs Bureau chief in April 1935 placed him in an increasingly favorable position to put his ideas about the national economy into practice. By 1936, Kishi became identified within his ministry and business circles not with industrial rationalization, but with the new trend of the "controlled economy."

Kishi continued to adhere to the basic principle behind the national economy—replacing unrestricted competition with cooperation in order to promote the welfare of the whole and national interests. What was new was the way in which "national interests" or "public welfare" was defined. Under the Hamaguchi and Wakatsuki cabinets, the national goal had been economic recovery through industrial rationalization. By thoroughly rationalizing the industrial process in terms of management methods, production technology, and worker efficiency, Japan would reduce its production costs, expand its overseas markets, and fully benefit from the devaluation of the yen following the reimposition of the gold embargo.[53] They viewed these measures as temporary but necessary measures to pull the country out of the depression. The short-term nature of these policies was reflected in the series of "temporary" organizations created to draft these policies, such as the Commerce and Industry Deliberation Council, the Emergency Industry Deliberation Council, and most important, the Temporary Industrial Rationality Bureau, as well as the five-year limit appended to the Important Industries Control Law.

Following the military's occupation of Manchuria, the general ideological shift to the right, and Japan's withdrawal from the League of Nations in 1933, however, the building of a national defense state became the state's goal. Now the national economy provided the framework for the army's program of total war mobilization, particularly heavy industrialization. Under the new national unity cabinets, state intervention not only continued but increased, despite the fact that the economy was well on its way to recovery. The 1933 amendment to and subsequent extension of the Important Industries Control Law, and the series of laws and drafts of laws providing for state intervention in the steel, oil, automobile, and electric power industries, suggested a new intent and direction

---

53. Kishi, "Sangyō gōrika yori tōsei keizai e," 10–11.

of state policy. Programs and policies pursued under the guiding spirit of national economy now came to be associated with the establishment of a controlled economy.

By 1934, Kishi began to emphasize the close connection between the movements for industrial rationalization and industrial control. As he explained:

> Today, in a somewhat revived industrial world, among those forgetful Japanese there seem to be many who view industrial rationalization as a subject of the distant past.... Some even cynically think that the word "controlled economy," which has been ringing loudly in our ears and about which I write here, has replaced it; and that the craze for "industrial rationalization" was abandoned and "controlled economy" became the new fad among us. In my opinion, however, this path from industrial rationalization to controlled economy cannot be explained by either of the two views that it was simply a fad or something arbitrarily devised by an eccentric to appeal to the public. I think that it lies in the fact that internally, industrial rationalization and controlled economy are intimately related, and consequently the progression from the industrial rationalization movement to the movement for a controlled economy is an inevitable trend.[54]

According to Kishi, the two movements shared the guiding spirit of the national economy, which he viewed as the "seed" that was planted via the industrial rationalization movement.[55] The controlled economy was the application and further elaboration of this spirit; in this sense, industrial rationalization represented the first step toward the creation of a controlled economy. As Kishi noted, the controlled economy could take one of two forms: a planned economy within a socialist economic organization as in the Soviet Union; or a controlled economy built on the foundations of the present capitalistic economic structure—which he preferred. Applying a Listian framework of economic stages of development, Kishi described the creation of the controlled economy as a natural progression within the historical development of the economy:

> When we think about the legal process of development of our economic livelihood from the past up until today, in the remote ages a self-sufficient economic unit was formed within the narrow sphere of the family unity. This economic unit then expanded, becoming a village economy, an urban economy or economy of a feudal state, and

54. Ibid., 2.
55. Ibid., 11, 25.

developed into today's national economy. Furthermore, it is said that this will probably develop into a bloc economy, or world or international economy. In any event, I think that there is no debate about the present age being the era of the national economy.[56]

Kishi pointed out that Germany and Italy were among the capitalist countries that had adopted a fairly advanced controlled economy. But another model closer to home was Manchukuo. The Kwantung Army adopted as its goal of economic construction "the attainment of the health and vitality of the entire national economy" or the creation of an "economy for the well-being of the people" (*anmin keizai*). They sought to realize this through its state-controlled special company system in the public service and heavy industrial sector. The idea of systematically targeting and nurturing key industries connected to national defense was applied in Japan in the form of a series of industry laws (*jigyō hō*) in strategic defense-related industries. The founding of Manchukuo and promotion of a Japan-Manchuria bloc removed inhibitions about implementing similar control techniques in Japan. The personal links among bureaucrats greatly facilitated this. Directing this effort in Japan were Yoshino and Kishi, and in Manchuria, Shiina Etsusaburō, Kishi's protégé and chief of the Control Section of Manchukuo's Industrial Department.

In the case of the petroleum industry, the creation of a special company in Manchuria and enactment of an industry law in Japan occurred almost simultaneously. In Manchuria, Shiina and his colleagues at the Industry Department established the Manchurian Petroleum Company (Manshū sekiyū kaisha) based on the army's plan for a special company in February 1934. Mantetsu was the major shareholder, financing 40 percent of the new company, with the remaining 60 percent financed in equal proportion by the Manchukuo government, Japanese zaibatsu interests such as Mitsui and Mitsubishi, and Japanese oil companies such as Nisseki and Kokura. Up until this point, Manchuria had obtained the bulk of its liquid fuel from foreign companies such as the American Standard Vacuum Oil Company and the Anglo-Dutch Asiatic Petroleum Company, and had imported only a small amount of fuel from Japan. They created the special company in order to develop Manchuria's oil fields, secure petroleum products through the processing of imported crude oil, and merge the activities of the existing oil concerns operating in Manchuria such as Mantetsu's oil-extraction activities at Fushun.

One month later, Japan adopted a similar strategy which took the form of the Petroleum Industry Law. Japan had also faced severe competition by the same

---

56. Ibid., 30–31.

foreign oil companies. In July 1931, under the Important Industries Control Law, the four major Japanese oil companies Nisseki, Kokura, Mitsui, and Mitsubishi had formed a cartel with two foreign companies to set uniform prices and assign market share. The oil cartel soon collapsed after the Soviet Union began to export cheaper oil. In the summer of 1933, the various ministries and the army and navy deliberated on a new petroleum policy and drafted a plan to promote the country's petroleum industry and private oil reserves, secure and develop oil reserves, and develop the synthetic oil industry. This plan became the basis for the Petroleum Industry Law. Similar to Manchuria's special company law, the new law required that oil companies obtain a government license and approval for their business plans, including any change, transfer, and suspension, or merger of its business. The government could intervene in the setting of price and quantity, and retained the right to purchase oil on demand. Unlike the special company law, however, the government did not interfere in the companies' equity structure, disposition of profits, or selection of personnel—reflecting the different ideological basis of Manchuria's partially planned economy and Japan's controlled economy. The result of both laws was the effective elimination of competition by foreign companies.

Before leaving for Manchuria, Kishi helped draft Japan's second industry law, the Automobile Manufacturing Law in May 1936. In terms of form and content, the law served as the model for subsequent industry laws in Japan.[57] It also served as a model for the Important Industries Control Law for Manchuria. The law required that automobile manufacturers obtain a license from the government. In order to qualify for a license the companies had to be majority owned by Japanese nationals and be engaged in the mass production of large vehicles. Like the Petroleum Industry Law, it sought to encourage the growth of Japan's automobile industry by providing major manufacturers with tax breaks, import duty exemptions, and subsidies on the one hand, while restricting foreign imports (in this case foreign automobiles and parts) on the other hand. Since the law was specifically created in the interests of national defense, it required that licensed firms provide automobiles and parts and specialized research and facilities for the military.[58] The two firms that received government licenses were Toyota and Nissan. Nissan's president Ayukawa Yoshisuke had wanted to import foreign capital and technology to compete with firms such as Ford and Volkswagen, but the army had opposed the idea. Eventually equipment was imported

---

57. Tsūshō sangyō sho, *Sangyō tōsei*, vol. 11, *Shōkō seisakushi* (Tokyo: Shōkō seisakushi kankō kai, 1964), 88.
    58. Ibid., 88–89.

and special licenses were granted.[59] Through the drafting of the law, Ayukawa formed valuable ties with Kishi and the army, which were instrumental in bringing Ayukawa and his entire Nissan operations over to Manchuria.

Clarifying the Concept of Control

Despite their power struggles with the Kwantung Army, bureaucrats served as its chief allies and lobbyists to the outside world. Bureaucrats softened and repackaged the army's blunt demands. The Kwantung Army was determined to exclude the evils of monopoly capitalism by imposing state control over industrial development via the special company system. But it also acknowledged that without the aid of private capital and technology, industrial development in Manchuria would be difficult. In its outline, the army welcomed foreign capital and technology as one of the state's four basic policies. Essentially, its message was that it welcomed the participation of private enterprise in Manchurian development, but only on the army's terms. These terms, however, were not specified, nor were the boundaries between controlled and freely managed enterprise clearly established. Not surprisingly, the outline and the "anti-zaibatsu" stance of the Kwantung Army did little to encourage private investment. Part of the problem was that private firms were hesitant to invest in Manchuria because of the unstable political situation and currency risks, particularly in the early period of military "pacification" and economic construction. The other reason, however, lay in the army's failure to establish the terms of control and explain the government's intentions behind its control policy, particularly with regard to private enterprise. One of the first tasks undertaken by bureaucrats was to address this latter issue.

Bureaucrats launched a public relations campaign to clarify the state's control policy through official policy statements, internal reports, and interviews with the press. These efforts culminated in the creation of Manchukuo's first control law in 1937. In this law, bureaucrats formulated the state's new "control ideology" (tōsei shisō), which presented the concept of "control" in a favorable light and sought to make it palatable to Japanese investors. Following the publication of the outline, the government issued a brief statement explaining its stance toward private enterprise in Manchuria entitled "Clarification Regarding General Enterprises" ("Ippan kigyō ni taisuru kōmei," June 1934). In this statement the government admitted that the outline had not clearly defined the spheres in

---

59. Kishi, Yatsugi, and Itō, eds. "Kankai seikai rokujūnen: Dai-ikkai Manshū jidai," 281.

which civilian enterprise could operate and specified those industries that would come under government control:

> With regard to industries necessary for national defense, public or public interest industries, and industries which are the foundation for other industries, namely transportation, communication, steel, light metals, gold, coal, petroleum, automobiles, sulphur, soda, and forestry, etc..., the government will devise special measures. With regard to other general businesses, a type of administrative control may be applied in accordance with the nature of the business, but in general, the government widely welcomes civilian participation and management."[60]

According to one internal report, "special measures" referred to management by the special companies. It claimed that the ideas of applying administrative controls and welcoming civilian participation and management were in no way contradictory but complementary policies.[61] Although the government was still vague about the meaning of "control" and about the specific industries to be controlled, it conveyed two important messages: first, that whatever "control" was, it would be applied selectively; and second, that the government's intention was not to reject free enterprise, but to welcome civilian enterprise, while applying certain forms of control in selected industries.

The following year in June 1935, bureaucrats clarified this policy in their "Request to Industrialists and Entrepreneurs" ("Kōgyō kigyōka ni taisuru yōbō"). After reiterating the government's policy put forth in the previous statement, it outlined the Business Department's provisional licensing procedures for private companies. According to this policy, companies were prohibited from establishing a private stock corporation and had to apply to the head of the Business Department for a license. In order to obtain a license, companies were required to submit information concerning their business activities and finances, as well as a business plan, the founder's résumé, and their policies for raising equity. The aim of the statement was twofold: to describe the licensing procedure for private businesses to the Japanese public, and to assure the public that the government's licensing requirements in no way represented an attempt to exclude private enterprise in Manchuria.[62]

---

60. "Ippan kigyō ni kansuru ken," contained in Kokumuin sōmuchō kikakusho, *Manshūkoku keizai kensetsu ni kansuru shiryō* (June 1, 1936), 41; see also: Furumi, *Wasureenu Manshūkoku,* 105–6; and Kimijima Kazuhiko, "Kōkōgyō shihai no tenkai," in Asada and Kobayashi, eds., *Nihon teikoku-shugi no Manshū shihai,* 572.

61. Manshūkoku jitsugyōbu tōseika, "Tōseihō seitei riyū" (September 16, 1935), in "Nichi-Man keizai tōsei hōsaku yōkō ni kansuru ken," Minobe Yōji Documents.

62. "Kōgyō kigyōka ni taisuru yōbō," in Manshūkoku kokumūin sōmūchō kikakusho, ed., *Manshūkoku keizai kensetsu ni kansuru shiryō* (Manshūkoku kokumūin sōmuchō kikakusho, 1937), 107–108.

Bureaucrats simultaneously began to lay the conceptual groundwork for the establishment of a control law for private enterprise in Manchuria. The law was the brainchild of Kishi and Shiina. As the main "idea man" and head of the Control Section at the Industrial Department, Shiina devised its basic concepts and framework.[63] Kishi, who assumed leadership of the department from October 1936, pushed for the law's passage.[64] In a classified document explaining the need for the law, bureaucrats conveyed the basic principles of state control.[65] Above all, state control aimed to "avoid the evils of capitalism of the past," especially "wasteful and lavish spending" (rōhi, ranpi), by investing Japanese and foreign capital in an efficient and prudent manner. Bureaucrats distinguished between two types of control: control in the negative sense of "suppression" (yokusei) and "obstruction" (bōshi), and control in the positive sense of "supervision" (kantoku) and "protection" (hogo). As the report explained, "Control does not mean obstructing and suppressing industry's rapid advance and inflow of capital, as is often mistakenly believed." Rather, through control, in which "supervision means protection" (kantoku wa hogo nari), the state provides a firm and stable basis for industry and facilitates investment in Manchuria. In addition, control referred to state supervision of important sectors of the economy for the purpose of achieving the comprehensive development of every sector of the economy and the efficient development of resources. Moreover, control referred to state management of those industries tied to the well-being of the people directly through public companies, or indirectly, through the special companies. It also referred to state "regulation" (chōsetsu) of all other industries, from both the aspects of production and consumption, in order to ensure the well-being of the people and support their livelihood. Finally, they equated control with fostering a close economic relationship of mutual dependence and coprosperity between Japan and Manchuria and the application of broad state controls to achieve that purpose.

Reform bureaucrats sought to explain these principles of the state's control policy to the Japanese public. In September 1936, Manchukuo's planners Hoshino, Shiina, Mōri, Furumi, Tamura Toshio, Takahashi, and Colonel Akinaga participated in a roundtable on Manchurian development sponsored by the *Oriental Economist* in Shinkyō. The purpose of the roundtable was to clarify Manchuria policy to the Japanese public, particularly with respect to the mainland's three primary concerns: the policy of excluding capitalists (shihonka hairu fuka),

---

63. Shiina Etsusaburō tsuitōroku kankōkai, ed., *Kiroku: Shiina Etsusaburō,* 124.
64. Furumi, *Wasureenu Manshūkoku,* 106.
65. Manshūkoku jitsugyōbu tōseika, "Tōseihō seitei riyū," Minobe Yōji Documents.

competition between mainland and Manchurian industries, and the negative effect of the special companies on the development of free enterprise.[66]

In their discussion, bureaucrats tried to convey their control policy in the positive sense of promoting economic development and ensuring the well-being of the people via state "guidance," "supervision," and "protection," rather than in the negative sense of discouraging zaibatsu participation and obstructing industrial advance in Manchuria and Japan. In addressing the concerns raised about the army's alleged policy of "prohibiting capitalists," the participants reinterpreted the policy in a more positive light as one of "avoiding the evils of capitalism." As Akinaga stated, "I don't know who said 'exclude capitalists,' ... if we are compelled to interpret the expression in our favor, the expression 'exclude capitalists' means the reform of the evils of capitalism, which is Manchuria's basic policy of economic construction." Hoshino then added:

> As for the policy of "excluding capitalists," if we elaborate Mr. Akinaga's point, it becomes two issues. One [issue] is, to put it simply, we are concerned that if we leave those industries which are the basis of the people's livelihood and the state economy to their natural course, they will be operated in the interests of each monopoly capitalist. This would cause great harm in aiming for the so-called development of the state of Manchukuo as a whole... one more [issue] is that, even with regard to the other industries, we will proceed, to the extent possible, without engaging in so-called "[activities] which possess a through and through, or pure and simple profit-making mentality without concern for anything else," or "[activities] which are unavoidable in the pursuit of profits, including those associated with deception and fraud."...In short, we will proceed by excluding from the start such things as the evils of capitalism or the system of capital which appear in Japan at present.[67]

Among the so-called "evils of capitalism" mentioned were "wasting capital and technology," labor conflict, and unemployment. As for the latter, Hoshino explained that "we are trying to avoid conflict between laborers and capitalists. Labor problems have become fierce as a result of pursuing things in a liberal, uncontrolled way. We are thinking of ways to prevent this type of situation." He noted that unemployment, which was the biggest social problem, could be alleviated through the state's control policy.

---

66. Akinaga, Hoshino, et al., "Manshūkoku keizai no genchi zadankai," *Tōyō keizai shinpō* (October 24, 1936), 27.

67. Ibid., 27, 30

With regard to the special company system, bureaucrats tried to convey the idea of state control in the sense of state "guidance" and "supervision." As Taka-hashi put it: "Basically, we make a general forecast of present and future demands and within this range have the factory equipment carry it out. Even in Japan, as you know, it appears that those with a lot of capital are doing this. Proceeding along this line, the method here is to adopt a so-called 'kind' approach toward industrialists—not just restraining them, but making them skillfully match sup-ply and demand and skillfully manage business with a kind heart." According to Akinaga, "So-called 'guiding supervision' (*shidō kantoku*) of the special compa-nies does not mean supervising and inspecting in the passive sense, but super-vising and guiding from the active standpoint of skillfully managing the special companies." Anticipating the Five Year Plan, bureaucrats referred to supervision and guidance of the special companies within the context of an overall plan. Tamura explained that

> since Manchukuo's special companies are responsible for the task of managing the important industries in a planlike way as a function of the state as a managed body (*keieitai*), the chief aim of supervision is to raise their efficiency while coordinating them within a comprehensive plan. It is guidance for the purpose of comprehensively bringing into full display the functions of every kind of special company as part of an overall plan.

Tamura used the case of state control of electric power as an example. He noted that "the liberal-type thinking is probably that electric fees become high when one company is given a monopoly"; whereas "in Manchukuo, because we proceed via a comprehensive control policy, monopolization does not lead to estrangement from the demands of the consumers."[68]

Bureaucrats tried to alleviate concerns about the potential competition be-tween mainland and Manchurian industries by emphasizing the mutually ben-eficial relationship that would result from the Japan-Manchuria bloc. Akinaga described this relationship as follows:

> I think that, as for the Japan-Manchuria one bloc, the economic goal must be to increasingly strengthen the actual comprehensive economic power of Japan and Manchuria as one body. Speaking from this stand-point, the idea that Manchukuo is a country of raw materials and the mainland is a country of processing industries, or the extreme main-land view that aims to protect existing industries and not develop

---

68. Ibid., 30, 31, 32.

Manchuria's industries too much, is wrong. In sum, we thought that it is not only fair and appropriate to aim to make it advantageous for both Japan and Manchuria based on the idea of "suitable place, suitable application" with Japan and Manchuria as one body, but it is also absolutely necessary in order for Japan to strengthen its economic capacity overseas in the future.[69]

At the same time, though, they hinted that for the Japan-Manchuria bloc to be realized, not only Manchuria, but Japan would have to be controlled.[70] Despite the rhetorical skills and public relations efforts of bureaucrats, private capital was slow to arrive. Not until Kishi wooed the new zaibatsu industrialist Ayukawa Yoshisuke with concrete and lucrative terms did Manchurian industrialization finally get under way.

Kishi Nobusuke and Manchurian Industrialization
Kishi's Arrival

Kishi's control orientation made him popular among army technocrats. The Kwantung Army had been seeking Kishi's expertise in Manchuria from the time it launched its industrialization program in the early 1930s. Having recently produced the first drafts for a Manchurian Five Year Plan in the fall of 1936, Manchukuo leaders now requested Kishi's assistance in its implementation. Kishi assumed control over the Business Department as director of its secretariat (*Jitsugyōbu sōmushichō*). In a major reorganization in 1937, Kishi assumed the roughly equivalent post of deputy head (*jichō*) of the renamed Industrial Department (Sangyōbu). From March 1939 onward, Kishi concurrently held the post of vice chief of the General Affairs Agency under Hoshino. During his three-year stay in Manchuria, he became one of the most powerful figures in Manchukuo.

As effective head of Manchukuo's Industrial Department, Kishi presided over the transformation of Manchuria from an independent command economy to a Japan-centered "national defense economy" (*kokubō keizai*). With the political and economic foundations of the new state established, Manchuria planners now embarked on the second stage of economic development, which the Kwantung Army laid out in its "Outline for the Second Stage of Construction of Manchukuo's Economy" ("Manshūkoku dainiki keizai kensetsu yōkō") of August 8,

69. Ibid., 33.
70. Ibid., 34.

1936.[71] Whereas the first outline had aimed at the creation of an independent national economy, the second outline aimed at the establishment of a national defense economy subordinate to Japan by 1940 or 1941. Now the goal became "the realization of facilities necessary for the joint defense of Japan and Manchukuo" and the "grounding of the roots of the imperial country's continental policy through the promotion of the healthy development of Manchukuo."[72] From the official perspective, the concepts of national economy and national defense economy were closely linked. As Manchukuo's official history explained: "It is the case that a national economy can be realized only via the national defense economy, and that we can expect for the first time the completion of a national defense economy based upon the foundations of the national economy."[73]

The army sought to attain self-sufficiency through the creation of an autarkic Japan-Manchuria bloc. The main vehicle was to be a Five Year Plan to develop Manchurian heavy and chemical industries. Under the new slogan of "onsite procurement" (genchi-chōbenshugi) and self-sufficiency (jikyū jisoku) in armaments, the army expected Manchuria to produce the necessary resources for its armaments industry as well as supply Japan with resources that it lacked.[74] As part of its plan for autarky, the outline called for policies to strengthen Manchukuo's agricultural base, which included a campaign to encourage Japanese immigration to Manchuria with the goal of relocating 1 million Japanese farmers in twenty years and 100,000 Japanese within the first five years.[75] At the same time immigration from Korea and North China was to be curtailed. The army also called for the streamlining of Manchukuo's administrative structure and the strengthening of state control through greater coordination between the center and regions and between the political and economic structure.

By the time Kishi arrived in Manchuria, the Five Year Plan had been more or less completed. Final drafting of the plan took place in Manchuria in October 1936 by a group of Kwantung Army officers, Mantetsu researchers, and bureaucrats, including Akinaga, Hoshino, Shiina, and Matsuda. The group established production targets for each sector based on a revised draft produced by Miyazaki Masayoshi. Following Ishiwara's departure, Miyazaki had returned to Japan and together with Ishiwara, established the Japan-Manchuria Economic Research

---

71. Kokumuin sōmuchō kikakusho, Manshūkoku keizai kensetsu ni kansuru shiryō (June 1, 1936), 57–63.

72. Ibid., 57.

73. Manshūkoku seifu, Meiji hyakunenshi sōsho: Manshū kenkoku, 311.

74. Hara Akira, "1930 nendai no Manshū keizai tōsei seisaku," 60; Fujiwara, Manshūkoku tōsei keizairon, 75.

75. For an examination of Japan's immigration policy toward Manchukuo, see Young, Japan's Total Empire, chapter 8.

Institute (Nichi-Man zaisei keizai kenkyūkai) in 1935 to assess Japan's defense capabilities in the event of war with America. In addition to the challenges of drafting the plan, the task of garnering support for the Soviet-style Five Year Plan within Japan presented formidable obstacles. The very idea of creating an "Industrial Gosplan" for Manchuria required some mental adjustment among Japanese bureaucrats and the business community. Even Kishi, who became one of Japan's foremost "control bureaucrats," recalled being taken aback at first hearing about the Soviet Union's own Five Year Plan:

> When I first learned about the plan, I was shocked to a certain degree. It was completely different from the liberal economy that we were accustomed to up until then and I remember feeling threatened by the type of thinking and determination to set targets and achieve them. However, I had my doubts as to whether they could really carry it out according to the plan. People like Mr. Yoshino [Shinji] were convinced that such a thing was impossible and wrote it off.[76]

One can only imagine how conservative bureaucrats at the ministries in Japan reacted, especially when presented with a request for the then unprecedented amount of 2.5 billion yen to implement the plan. According to Furumi, when General Affairs Agency chief Hoshino, Kishi, and other Manchuria planners traveled to Tokyo to request funds for the plan in December 1936, the Finance Ministry's initial response was, "Manchukuo has brought an outrageous plan. We won't have anything to do with it."[77] Likewise, the ministries of commerce and agriculture fiercely opposed the plan on the grounds that it conflicted with the interests of Manchurian and Japanese industries.[78] From December 1936 until early 1937 Manchukuo bureaucrats, Mantetsu researchers, and Kwantung Army officers worked with the Army Ministry to revise the plan and obtain the approval of the various ministries in Japan. Deliberation and revisions of the plan took place at the Manchurian Affairs Bureau, which was the only agency to fully endorse the plan before its launching in April 1937.

Following the outbreak of full-scale war with China, however, the Japanese government quickly reversed its cool attitude and fully embraced the Five Year Plan. Now the government demanded that Manchurian planners expand the scale of production and shorten the plan's timetable. According to the historian

---

76. Kishi, Yatsugi, and Itō, "Shōkō daijin kara haisen e," 283.
77. Furumi, *Wasureenu Manshūkoku,* 115.
78. Hara Akira, "'Manshū' ni okeru keizai tōsei seisaku no tenkai: Mantetsu kaisō to Mangyō sestsuritsu o megutte," in Andō Yoshio, ed., *Nihon keizai seisakushi ron* (Tokyo: Tokyo daigaku shuppankai, 1976), 229–230.

Hara Akira, production targets for resources for munitions production were dramatically increased. Targets for coal, electric power, and vehicles were increased 80 percent, while those for commodities such as aluminum, magnesium and metals were roughly doubled. The new estimated cost of the plan was more than doubled to 5 billion yen.[79] The China War was a shot in the arm for Manchuria planners, so much so that after the war, prosecutors at the International Military Tribunal for the Far East raised the issue of whether Manchuria planners and Japanese troops in China had conspired to use the incident as a way to obtain Japan's support for the Five Year Plan.[80]

Delimiting Control

Although Kishi embraced the army's project to create a national defense economy, he had markedly different ideas about how to realize the army's goals. Kishi and his colleagues sought to win private industry over to the state's goals of national defense and incorporate the benefits of free enterprise—management expertise, capital, and technology—into a planned economy. One strategy, which they had been pursuing since their arrival, was to carve out a sphere for private enterprise within Manchuria's economy. In order to alleviate the concerns of Japanese business about investing in Manchuria, Kishi pushed through the state's first control law for private enterprise in Manchuria, the Important Industries Control Law, in May 1937.

While its immediate aim was to set down government regulations for the so-called "semispecial companies" and licensed private companies which fell outside of the government's special company laws, the new law revealed a broader political intent. First, Manchukuo bureaucrats sought to mark out the boundaries of control. In one official statement, the government explained its reasons for drafting the law. Up until then, the state's industrial policies had been comprised of emergency measures that had not been set down as official laws and regulations. These measures had caused considerable misunderstanding about the purpose, scope, and degree of the government's control. As a result, the government feared that these policies might create a barrier to stable and healthy business activity. Especially with the recent abolition of extraterritoriality and the transfer of the former adjoining territories of Mantetsu to the state, the government felt it necessary to establish general laws to clarify the contents of control and

---

79. Ibid.
80. International Military Tribunal for the Far East, *Transcript of Proceedings: General Index of the Record of the Defense Case through the Tri-Partite Pact Section of the Pacific Phase* (Tokyo, 1947), 20, 433.

provide clear directives for business activities.[81] Bureaucrats described the new law and the Five Year Plan as representing "two sides of a coin."[82] According to Kishi, they passed the law at the time of the launching of the Five Year Plan in order to show that the government made a distinction between those industries controlled under the special company law, those regulated under the new Important Industries Control Law, and those left completely to private management.[83] In other words, the government did not seek to institute control "for control's sake," or for the purpose of creating a socialist-type economy. At the same time, the government sought to convey the message that all industries, not only the special companies, were expected to participate in the Five Year Plan. As Kishi explained, "Furthermore, this industrial control seeks to develop the national economy based on totalistic goals, to actively embody the planned economy, and aims for Manchuria, a "'have' country" (*moteru kuni*), to bypass the stage of light industry and advance directly to the stage of heavy industry."[84]

As with Japan's own Important Industries Control Law of 1931, the law designated twenty-one industries as "important industries" that came under state control and specified how these industries were to be controlled. Among the twenty industries designated were those related to munitions, aircraft, automobiles, liquid fuel, iron, steel, coal mining, textiles, cement, and fertilizer. Within these industries, the government specified which types of products were to be controlled. According to the law, firms engaged in important industries were to operate under a license by the Supervisory Board Minister, to whom these firms would present all business plans and reports. When the state considered it necessary, it could order these firms to report the state of their business and financial condition and to inspect the firms' accounts and deposits. Permission of the Supervisory Board Minister was required in the case of enactment or revisions of control agreements, expansion or changes in production facilities, transfer of business units, and mergers. Firms were also ordered to report any suspension or discontinuation of lines of business or dissolution of a corporation.[85] In addition, in the case of violations of the law, penalties in the form of fines of up to five thousand yen could be imposed.

Second, in a statement issued by the Business Department, bureaucrats officially set out the reasons for control. The statement explained that the

---

81. "Jūyō sangyō tōsei hō shikō ni kansuru ken: Riyūsho," July 27, 1937, contained in Fujiwara, *Manshūkoku tōsei keizairon*, 415.

82. Shiina, *Kiroku: Shiina Etsusaburō*, 124.

83. Kishi, "Sangyō kaihatsu gokanen keikaku no kangaekata" (1939), in Manshū kaikoshū kankōkai, *Aa, Manshū* (Tokyo: Nōrin shuppan kabushikigaisha, 1965), 239.

84. Ibid.

85. Fujiwara, *Manshūkoku tōsei keizairon*, 408–409.

government's industrial program aimed at the national goal of establishing a fused Japan-Manchuria economy and through it, "outwardly plan for the strengthening of broad national defense, and inwardly look to the stabilization and advancement of a national economy."[86] Based on the principle of "suitable site, suitable applicability" (*tekichi-tekiō shugi*), so-called "important industries" were to be targeted and their development coordinated between the two countries in order to establish and strengthen the Japan-Manchuria economy as a whole. The statement also explained that the law principally targeted the manufacturing and mining companies, that is, industries connected to the Five Year Plan. As for industries outside of the twenty industries, the government would "entrust its economic establishment and development completely to the initiative of entrepreneurs."[87] The government also differentiated between its policies toward the special companies and its policies toward other companies. The special companies were established based on the idea of "one industry, one company" and their internal management was placed under special state guidance and supervision. In contrast, the Important Industries Control Law placed limits on state control of the semispecial and licensed companies and sought to control only the external aspects of the firm's activities.[88] In other words, while the state sought to control certain external matters concerning mergers and changes in business lines that might affect other firms and industries, it would not intervene in such internal matters as the equity structure of the firm, the appointment and dismissal of managers, and the disposition of profits and dividends. In this way, bureaucrats sought to convey Manchukuo's control policy in the positive sense of actively promoting economic development through state assistance and guidance of firms as in Japan's control law. As one drafter put it, "Even if we say 'control' it is one that incorporates not a repressive policy (*yokuseisaku*), but an activist policy (*sekkyokusaku*) for the purpose of developing capital and enterprise."[89]

Having established the government's policy toward the semispecial and licensed companies, Kishi and his colleagues now turned their attention to the special company system. According to Kishi, the problem with the special companies was not a lack of capital, since Japanese government funds were available, but business expertise. He recalled saying at the time that "in order to realize the Five Year Plan, we do not need Mitsui and Mitsubishi's capital,

86. "Jūyō sangyō tōsei hō shikō ni kansuru ken: Jitsugyō bu kōmeisho," contained in ibid., 416.

87. Ibid.; Kimijima Kazuhiko, "Kōkōgyō shihai no tenkai," in Asada Kyōji and Kobayashi Hideo, eds., *Nihon teikokushugi no Manshū shihai: Jūgonen sensōki o chūshin ni* (Tokyo: Jichōsha, 1986), 572.

88. Fujiwara, *Manshūkoku tōsei keizairon*, 417–418.

89. Ibid.

what we want is their management expertise.... There are no capable managers in Manchuria to realize the Five Year Plan. They are all retired civil servants, and the leaders of the army faction that staged the Manchuria Incident also lack management ability." Kishi and others exerted much effort to bring Mitsui and Mitsubishi over to Manchuria, but without success. Manchurian development became possible only after Ayukawa moved over his entire Nissan management force.[90]

Before Kishi's arrival, the special company system had primarily reflected the concept of control of Kwantung Army officers and Mantetsu researchers. Army planners interpreted "control" as, above all, preventing zaibatsu domination of the economy through the creation of government monopolies in important industries. They believed that in order to promote the "rational" development of industry to further the army's defense goals, the special companies were to possess a "national planning character" and be large, comprehensive, and rapidly developed. Moreover, a form of administrative control or "supervisory guidance" (shidō kantoku) would be applied to achieve a balance between direct state control of the socialist systems and regulatory control of the capitalist systems.

The case of Manchuria's coal industry revealed the weakness of the army's strategy. In 1934 the Kwantung Army and Mantetsu created the Manchuria Coal Mining Company as a special company for the production of coal. Previously Mantetsu had controlled the majority of coal production in Manchuria through its mines, which included its famous Fushun mine.[91] At the time of the founding of the company, the Manchukuo government and Mantetsu each provided half of the initial capital of 16 million yen. The first director was the former Mantetsu president and Economic Research Association research head Sōgō Shinji. He was succeeded by Kōmoto Daisaku, the officer who helped stage the assassination of Zhang Zuolin. As president of the Manchurian Coal Company, Kōmoto gradually acquired most of the coal mines, except for Mantetsu's mines. He was regarded as the "king of coal" in Manchuria.

Reflecting the army's policy of "one industry, one company," the company, together with the Mantetsu's Fushun mines, held a monopoly on coal production in Manchuria. The company certainly possessed a "national planning character" since it was an official state organ funded by the government and Mantetsu and geared toward achieving the Five Year Plan targets. But the company's activities hardly represented a "rational distribution" of industry. Mantetsu retention of

90. Kishi, Yatsugi, and Itō, "Shōkō daijin kara haisen e," 287.
91. The above discussion is largely based on Hara Akira, "'Manshū' ni okeru keizai tōsei seisaku no tenkai," in Andō Yoshio, ed., Nihon keizai seisakushi ron (Tokyo: Tokyo daigaku shuppankai, 1976), 257–264.

the Fushun mines reflected political turf battles. Mantetsu fought hard to retain its Fushun mines even though it was unrelated to its main business of railway management and succeeded in retaining control over the mine when Mangyō took over the Manchurian Coal Company in 1937. The large number of acquisitions between 1934 and 1938, and expansion of its capital base to 80 million yen in 1937, demonstrated that the company was big, comprehensive, and rapidly developing. Mangyō, however, viewed the company's rapid expansion and unwieldy size as reckless. Despite large infusions of capital in 1937, the company managed to produce only roughly two-thirds of the Five Year Plan's projected amount in the first three years, in contrast to Mantetsu's Fushun mines and Okura zaibastu's Penhsihu Iron Works, which met the plan's production targets. The Manchurian Coal Company's performance paled in comparison to that of the privately managed coal companies in Japan.[92]

With deliberations for an ambitious Five Year Plan underway, the Kwantung Army was anxious to place the special companies on a firm footing. For the army, the main problem with the special companies stemmed from the monopolistic position they held in their respective industries. Although the army sought to avoid the problems associated with state-owned enterprises by creating joint stock companies, it essentially faced similar problems: the strong bureaucratic character of management, lack of capable personnel with sufficient business experience, low efficiency, and lack of incentive to improve performance.

In an attempt to eliminate the abuses inherent in the monopoly system, the army proposed measures to reform the special companies, which it outlined in its "Policy of Supervisory Guidance of Manchukuo's Special and Semispecial Companies" (July 1936). The army tried to strengthen state control over the special companies by expanding and strengthening the government's supervisory organ, adopting measures to actively supervise and control the special companies. The state proposed recruitment of persons with sufficient experience and vision. As for its control policy, the army sought to nurture the ability of managers and eliminate the "prosecutorial attitude" and interference in company management. This would be achieved by reducing the number of directors in each company, rationalizing the compensation system by controlling and gradually reducing the salaries of personnel, and restructuring the company's organization. In order to eliminate the abuses arising from the company's monopoly of industry, the army recommended the state's careful investigation of each company's business, the creation of new business plans to fit the present situation, the establishment of a balanced and equitable standard

---

92. Ibid., 258–263.

for the disposition of profits of each company.[93] The army sought to fine-tune the existing system by strengthening and centralizing control at the state planning level and curtailing bureaucratic interference in daily management at the enterprise level.

In contrast to the Kwantung Army's proposal for piecemeal reforms within the existing system, Kishi aimed to completely overhaul the special company system by transferring Ayukawa's Nissan operations to Manchuria. The events that led to the decision to bring Ayukawa and his company to Manchuria were set in motion in the fall of 1936, when the Kwantung Army invited heads of the old and new zaibatsu to visit Manchuria. Within this group, Ayukawa expressed enthusiasm for Manchukuo's industrial program and had attracted the attention of Ishiwara. Impressed with Ayukawa's management vision and plan to bring in foreign capital, Ishiwara and others at army headquarters decided to invite him to come to Manchuria to develop heavy industry. From the Manchuria side, Hoshino and Kishi led the efforts to bring over Ayukawa. Both made separate visits to Tokyo to recruit him. At the same time, in Manchuria they persuaded those Kwantung Army leaders and Manchukuo bureaucrats, such as Commander in Chief Ueda, Chief of General Staff Tōjō, Shiina, and Matsuda, to support Ayukawa's move to Manchuria. Senior Kwantung Army leaders were reluctant to change their "one industry, one company" policy and allow only Nissan to profit. They also feared that the deal would further strain their relationship with Mantetsu.

Kishi was pivotal in bringing over Ayukawa. Both were from Yamaguchi prefecture and had attended the same high school. Kishi got to know Ayukawa while drafting the Automobile Manufacturing Law. This was the beginning of a close relationship between them spanning over thirty years.[94] In July 1937, Kishi traveled to Tokyo to work out an agreement with Ayukawa for the establishment of Mangyō, a new Manchurian government–private business consortium that would change the course of economic development in Manchuria. Nissan's entire operations would be transferred to Manchuria to form the basis of the new company. It would acquire the government's special companies and semispecial companies owned by Mantetsu and would be managed by Ayukawa and his employees.

---

93. Kantōgun shireibu, "Manshūkoku tokushu kaisha oyobi juntokushu kaisha no shidō kantoku hōsaku" (July 24, 1936), in Manshūkoku kokumuin sōmuchō kikakusho, ed., *Manshūkoku keizai kensetsu ni kansuru shiryō* (Shinkyō, Manchuria: Manshūkoku kokumuin sōmuchō kikakusho, 1937), 55–56.

94. Ayukawa Yoshisuke sensei tsuisōroku hensan kankōkai, *Ayukawa Yoshisuke sensei tsuisōroku* (Ayukawa Yoshisuke sensei tsuisōroku hensan kankōkai, 1968), 1.

Mangyō

Through the creation of Mangyō, bureaucrats implemented their idea of control. In contrast to the Kwantung Army's negative approach to control of preventing zaibatsu domination, restricting profits, reducing compensation, and imposing more state controls, bureaucrats sought to apply control in the "positive" sense of "an active policy for the purpose of developing capital and enterprise." They did this by creating incentives to improve performance and minimize the financial risks to Nissan. In contrast to the Kwantung Army's strategy of restricting profits by placing caps on company dividends, the Manchukuo government guaranteed Nissan shareholders 6 percent dividends on their common shares for ten years. Any excess profits would be paid out to Nissan shareholders and the Manchukuo government in the ratio of two to one, with no overall limit on dividend payout rates. These terms represented an important departure from the army's policy of restricting the "limitless issue of dividends" of the special companies. Likewise, in the case of dissolution of the company, its assets would be split two to one between Nissan shareholders and the Manchukuo government. The government also guaranteed Nissan the principal and a 6 percent return on its investments for ten years—which meant that Mangyō debt was guaranteed by the Manchurian government.[95] In addition, the company received preferential tax treatment for ten years, and its share price was supported by the government in the stock market. The generous terms offered to Nissan was an acknowledgement by the government that the profit motive was a central motivating power of the firm and had to be incorporated to some extent into the special company system.

In establishing Mangyō, leaders took the position that pursuing profits and furthering the state's defense goals were not necessarily incompatible. The special companies could pursue profits and possess a national character, be large and comprehensive, and be rapidly developed. As a special company, Mangyō was financed by the Manchukuo government and Nissan on a fifty-fifty basis, with each side putting up 450 million yen. Bureaucrats argued that Mangyō was established not for the pursuit of profit, although profit incentives were provided to Nissan, but to achieve the state's defense goals of establishing heavy industry in Manchuria. As for the head of the new company, Ayukawa represented a compromise to the anti-zaibatsu stance of the state's founders. As a new zaibatsu leader, Ayukawa could be portrayed as a new type of industrialist and his company as a "revision of zaibatsu capitalism," based on its form of a publicly owned holding company. Not all were convinced by this argument, however. Some members of the Concordia Association viewed Ayukawa's arrival and the

---

95. Fujiwara, *Manshūkoku tōsei keizai ron,* 196.

creation of Mangyō as a "one hundred eighty degree reversal" of the anticapitalist policy of the founding faction.[96]

Under the new management structure, the principles of the multilateral organization of new technology-based industries were applied to the national economy as a whole. Ayukawa developed heavy industry on an even larger scale and in an even more comprehensive manner. Mangyō consolidated the steel, light metals, automobile and aircraft, and mining industries. In his report to the army, Ayukawa had criticized the basic special company system for pursuing what he described as the "line method," which he contrasted with the more preferable "pyramid method." For Ayukawa, the line method was based on the principle of "one industry, one company," in which industries were established in a linear, one-dimensional fashion in which each product was viewed in isolation: as Ayukawa put it, "a car is a car, a plane is a plane, coal is coal." As a result, the various industries were established in a random fashion without any logical connection to each other. Ayukawa also criticized Manchuria's special and semispecial companies as developing into subordinate "branches" (*shiten*) of Japan's own industries, having been built on the small-scale of Japanese industries as well as having become dependent on mainland businesses for both funds and "junior" personnel. Given this approach, "it was no wonder that Manchurian industry has been delayed."[97] Instead Ayukawa proposed the adoption of the "pyramid method" based on a new set of principles of development. Resources should be developed across the board in an integrated, planlike, comprehensive manner, and industries organized in a pyramid structure based on the principle of "many industries, one company" (*sūgyō issha shugi*) or "totalism" (*zentaishugi*). As Ayukawa explained, "If control is implemented based on 'totalism,' so called comprehensiveness (*sōgō*) will be achieved for the first time."[98] "We must completely extricate ourselves from the Japanese way of thinking and devise a new method geared toward Manchuria itself." Ayukawa argued that in a vast place like Manchuria the so-called "branch-scale" of industry was inappropriate; plans needed to be implemented on an immense scale. Ayukawa saw the need for what he referred to as a "pioneer mentality" similar to that in America or the Soviet Union. He viewed the Soviet Union's planned economy or America's Ford method of production as examples of applying deductive thinking to develop industries based on the natural conditions of that particular country or region.

---

96. Koyama, *Manshū kyōwakai no hattatsu*, 44.
97. Ayukawa Yoshisuke, "'Nissan' Manshū ichū e no hōfu—Watashi no yarō to suru Manshū sangyō kaihatsu shin soshiki" (interview), *Tōyō keizai shinpō* (November 1937), 27.
98. Ibid., 28.

Under Ayukawa, industry was developed even more rapidly than under the previous system. One of the main reasons the army was interested in Ayukawa was because of his plan to bring in foreign capital and technology to Manchuria. In a report to Ishiwara, Ayukawa put forth a plan to fulfill the army's ambitious production targets in five years. Under the current plan, Ayukawa argued that five years was unrealistic and would probably require a minimum of ten years, or possibly twenty years if Mantetsu tried to carry out the plan itself and proceeded rashly.[99] He believed that the military's goal of five years would be possible if Japan obtained 1 billion yen in foreign funds, of which a large portion would come from America. Of the 1 billion yen in foreign funds, Ayukawa estimated that, at the exchange rate of four yen to the dollar, Japan could borrow $250 million from America, with most of these funds going toward the purchase of the most modern machinery and tools from American companies such as Ford. Japan should be prepared to offer a split in profits from Manchurian heavy industry with American investors on a fifty-fifty basis in order to obtain American financing and equipment.[100]

Finally, leaders argued that industrial development under Mangyō represented a more rational organization of industry than under the previous system. Under the old system, the Kwantung Army had used Mantetsu as the principle vehicle for private investment and development of heavy industry. Through the establishment of Mangyō, technocrats were able to effectively curtail Mantetsu's participation in the Manchurian industrial program. Mantetsu, which had held a major stake in the special companies, did not have a direct financial stake in Mangyō. Mantetsu was forced to spin off all of its major heavy industrial concerns such as Shōwa Steel Works, Manchurian Light Metals, and Dōwa Motors, with the exception of its Fushun coal mines. Based on a 1937 plan, Mantetsu was to sell sixty-eight companies to the government for a sum of 550 million yen; these companies were then placed under Mangyō's management.

Participation in the Manchurian venture was a transformative experience for a select group of bureaucrats. They arrived in Manchuria as young, idealistic bureaucrats who had been educated in the rarefied and sheltered environment of Tokyo Imperial University and groomed for elite careers in Japan's ministries and business. Responding to calls to serve the nation, many left families

---

99. Ayukawa Yoshisuke, "Manshū jūkōgyō kaihatsu kaisha no hanashi," in Manshū kaikoshū kankōkai, ed. *Aa Manshū: Kunitsukuri sangyō keihatsusha no shuki*, 243.

100. Ibid., 244.

behind and devoted a good part of their youth in an extremely isolated place without the comforts of the homeland. They brought to their new posts a vision of modernizing a backward land by developing its economy and introducing modern efficient administrative techniques; they left with a pioneering mission to reform Japan.

These bureaucrats adapted to their new environment. They mingled with an eclectic community of military officers, right-wing activists, and left-wing planners who flourished in Manchuria's unrestricted environment. More important, they formed long-term working relationships with other elite military and civilian technocrats. They actively promoted their own managerial visions by recasting the ideologically rigid programs of Manchukuo's founders into more viable business ventures and devised bold and innovative plans to transform Manchukuo into a techno-fascist state. Their Manchurian experience was perceived as such a success that by 1940 Manchukuo's expanding bureaucracy became the training ground for elite bureaucrats as well as a haven for those seeking an alternative to the highly factionalized bureaucracy in Japan.[101] Shiina, who served as minister of the powerful Ministry of International Trade and Industry after the war, later recalled how Manchuria was the "great experimental ground" (*dai jikken jo*) for Japanese industry.[102] In Manchuria, technocrats laid the foundations for Japanese heavy industry and technocratic control and provided a model for Japan's wartime and postwar economy.

101. Yamamuro, "'Manshūkoku' tōchi kateiron," 117–111; Kan, interview, 246.
102. Shiina Etsusaburō, "Nihon sangyō no dai jikken jo—Manshū," *Bungei shunjū* (February 1976), 106–114.

IDEOLOGUES OF FASCISM

Okumura Kiwao and Mōri Hideoto

Reform bureaucrats returned from their overseas postings in Manchuria and China with a new mandate to reform Japan, Manchuria-style. Under Prime Minister Konoe, who headed three cabinets between 1937 and 1941, they assumed key bureaucratic posts. Many joined the newly established Cabinet Planning Board and held joint appointments at their old ministries. These bureaucrats drew on the ideas and support of civilian technocrats in reformist bodies such as the Shōwa Research Association, National Policy Research Association, and Social Masses Party. They attended informal, weekly discussion groups of military and civilian technocrats to deliberate on national policy. Through their writings, speeches, and interviews, they promoted a new techno-fascist vision of state, society, and empire.

Both the military and civilian programs promoted antiliberal visions and policies, but there were important differences. Reform bureaucrats and their supporters respected the Meiji constitution and preferred to change the system from within. They wanted to strengthen the state's control over society but rejected notions of nationalizing industries, eliminating the peerage system, or limiting private property. In their vision of the new order, ownership of property was not a requirement for its control. Moreover, they incorporated a new technological worldview and managerial and ethnic nationalist concepts drawn from European fascism into the military's vision. Finally, their techno-fascist vision was oriented primarily toward progressive, middle class, urban professionals.

Scholars of Japanese fascism have tended to emphasize the nativist, agrarian, and pan-Asianist visions associated with right-wing ideologues and radical officers. Maruyama Masao argued that the distinctive features of Japan's fascist ideology were the Japanese family system, physiocratic celebration of Japanese agriculture and village life, and pan-Asianism. He attributed these characteristics to the social basis of fascism, which he located in the provinces among the "petty bourgeois" class of small landowners, independent farmers, local officials, primary school teachers, and small business owners. Maruyama distinguished this group from the urban, salaried middle class, and especially Japanese intellectuals, who he claimed were too sophisticated and cosmopolitan to stoop to such a low cultural level.[1] A number of studies have challenged this view and have shown how elite, progressive intellectuals and left-wing politicians counted among fascism's supporters.[2]

Progressive intellectuals and politicians formed part of a broader group of university-educated, middle-class professionals who represented Japan's new managerial class. They included company managers, scientists, engineers, technicians, journalists, radio broadcasters, social scientists, and civil servants. These professionals desired neither capitalism nor socialism.[3] They preferred to eliminate the privileges of the upper class, alleviate the poverty of the working class, and increase their own power and prestige as administrators within the modern, complex organizations. At the same time, the rise of the managerial class altered the old political landscape of capitalists versus workers. Left-wing, progressive-minded intellectuals and politicians "graduated" from socialism and lost their revolutionary fervor as leaders of labor.[4] Under the new order, future political battles would take place between the "inner" and "outer" group of administrators.[5] Given the prestige of "insiders," or officials, in Japan, however, "outsiders" such as intellectuals and politicians would continue to play second fiddle to bureaucrats.

In order to fully grasp the modern, technocratic thrust of Japanese fascism, one must examine the ideology of two of its leading ideologues Mōri Hideoto and Okumura Kiwao. English language studies have analyzed the ideas and strategies

1. Maruyama, "The Ideology and Dynamics of Japanese Fascism," 36–51, 57–59.
2. James Crowley, "Intellectuals as Visionaries of the New Asian Order," in James W. Morley, ed., *Dilemmas of Growth in Prewar Japan* (Princeton: Princeton University Press, 1971), 319–373; Fletcher, *Search For a New Order*; Andrew E. Barshay, *The State and Intellectual in Imperial Japan* (Berkeley: University of California Press, 1988); William D. Wray, "Asō Hisashi and the Search for Renovation in the 1930s," *Papers on Japan* 5 (Cambridge, MA: East Asia Research Center, Harvard University, 1970), 55–98; Earl H. Kinmonth, "The Mouse that Roared: Saitō Takao, Conservative Critic of Japan's 'Holy War' in China," *Journal of Japanese Studies* 25, no. 2 (summer 1999), 331–360.
3. See George Orwell, "Second Thoughts on James Burnham," *Polemic* 3 (May 1946).
4. This point was made about postwar managerial groups by Walter A. Weisskopf, "Same Old New Class," *The New York Review of Books* 9, no. 10 (December 7, 1967).
5. Ibid.

of civilian right-wing ideologues and progressive intellectuals. We know com-
paratively little about the ideas and strategies of bureaucrats.[6] Political insiders
described Mōri as the "idea man" at the Cabinet Planning Board and the "bright
star" at the Asia Development Board. Working behind the scenes, Mōri served as
a central link between military officers, bureaucrats, pan-Asianists, and left-wing
intellectuals. Okumura assumed a more public role, first as author of the controversial Electric Power Control Law, and later as deputy chief of propaganda. He
delivered the regular radio broadcasts to the nation at the height of the Pacific
War. This chapter examines the ideas of Okumura and Mōri. I examine how they
combined technocratic planning with the Japanese *Volk* (*minzoku*) to formulate
their techno-fascist vision of Japan and its Asian empire. As will be shown, their
technocratic embrace of fascism did not represent a sudden, radical conversion
(*tenkō*) or reactionary turn to the past, but rather a gradual and considered embrace of new political trends and modern technology.

Combining Technology and *Volk*

Toward a Technological Worldview

Reform bureaucrats and their colleagues believed that technology was bringing
about not only the technicization of industry, war, and administration, but also
the technicization of the world order. As they saw it, the old liberal world order
based on the control of natural resources was being replaced by a new fascist
world order based on science and technology. Under the liberal world order,
Japan and Germany were "have-not" countries because they lacked the ideal balance of natural resources, markets, and capital to minimize economic instability
and provide economic self-sufficiency. Their weaknesses derived not so much
from late industrialization but from their lack of natural resources. Moreover,
their industries were unable to operate at their optimal levels because they lacked
sufficient raw materials, capital, and markets. In the case of Germany, the historian Robert Brady remarked that the "supreme paradox" of German industrial
rationalization was that it needed to be carried out within an organizational
sphere or unit that was greater than Germany itself.[7] Japanese leaders believed

---

6. On Okumura, see Gregory Kasza, "Fascism From Above? Japan's Kakushin Right in Comparative Perspective," in Stein, Ugelvik, and Larsen, ed., *Fascism outside Europe: The European Impulse
against Domestic Conditions in the Diffusion of Global Fascism* (Boulder: Social Science Monographs,
2001), 183–232.

7. Brady, The *Rationalization Movement in German Industry,* 324.

that in order for their country to become a world power, it needed technology, national spirit, and a regional bloc.

In their techno-fascist worldview, the key to Japanese hegemony in Asia was Japan's liberation from natural resources. If "have-not" countries could overcome their resource limitations by creating synthetic substitutes, their entry to the exclusive circle of resource-rich powers would no longer be blocked. This lesson was poignantly brought home early on by Kishi Nobusuke. After observing the state of American industry in 1926, he reported the following impressions:

> The United States was a country blessed with the power of its vast richness—namely its abundant natural resources, enormous capital base, and lively spirit of enterprise. Its vast economic structure was unimaginable to the Japanese and made Japan's own economy appear extremely shabby in comparison. For instance, the numerous oil well towers in the area surrounding Los Angeles appeared as a forest from a distance. Several days worth of production from that area exceeded the amount Japan could produce in one year. Likewise in the case of iron production, whereas the United States was producing two million tons in one month, Japan's own Yawata Steel was struggling to achieve its target of one million tons in several years."[8]

Concluding that Japan's economic potential could not possibly ever match that of the United States, he set off gloomily on his return trip via Europe. Kishi recalled how his spirits improved considerably when he visited Germany. He was deeply impressed by the spirit of cooperative effort in Germany's industrial rationalization movement and the strategic approach toward technology, especially the application of new chemical technology to create synthetic substitutes for natural resources that it lacked. Kishi perceived a solution to Japan's own resource problem. By directing scientific and technical research toward areas considered vital to the state, technologists could devise creative solutions to Japan's problems. Upon his return, Kishi became a leading proponent of the cooperative spirit and science and technology.

A similar view informed the philosophy of Riken. Ōkōchi Masatoshi cautioned that Japan needed to pursue its own path and not simply follow the Western trajectory of science and technology. The key to Japan's future, he asserted, was the creation of high quality, indigenous products based on science. He rejected the liberal perception of Japan as a "have-not" country because it interpreted resources in terms of Western needs. For Ōkōchi, resources were something that a

---

8. Kishi, "Sangyō gōrika yori tōsei keizai e," 3–5.

country creates, not possesses.[9] He pointed out the example of aluminum production. Japan faced considerable challenges because it lacked the key resource, bauxite. Since Western countries possessed bauxite, they never considered using alternative resources to produce aluminum. Japan, on the other hand, compensated for its deficiencies in bauxite by devising a way to produce aluminum using indigenous resources such as "miso" and kanuma clay and alum.[10] Ōkōchi had faith in Japan's ability to develop its own scientific and technical knowledge and new products using the rich material and human resources in Japan and Asia. Ōkōchi did not believe in science for science's sake, but rather science for the purpose of creating value and enhancing Japan's competitiveness.

The Social Masses Party leader Kamei Kan'ichirō was a forceful advocate of the Nazi technological spirit. After returning from a visit to Nazi Germany in 1938, Kamei began to promote its technological worldview. Kamei explained that the Nazis attributed the collapse of internationalism not to the rise of nationalism but to the implosion of the old order. In their view, the liberal capitalist system lost its ability to promote the welfare and culture of mankind as a result of several factors, among which was the failure to recognize the fundamental transition in the world order from one organized by the resource-based international division of labor to one organized on the basis of production facilities and technology. With regard to the latter, Kamei quoted the Nazi economics minister Walter Funk, who boasted about Germany's recent technological advances resulting from the widespread use of electricity from the late 1880s and particularly in the last decade:

> Specifically, in terms of their practical application, the rapid development of electromagnetic theory and the development of electron theory within theoretical physics promoted a series of developments in the precision machine and machine tool industries and wire and wireless precision electrical machinery and electrical meter industries. Furthermore, as a raw material source, electricity gave rise to the new light alloy industries and promoted a line of synthetic chemical industries. On the other hand, the gauge bloc and limited gauge operation as well as the precision machine tool industry and Zeiss Ikon spectroscope were endorsed, and the machine industry's precision specification of two to three millimeters became the norm.[11]

9. Ōkōchi kinenkaihen, ed., *Ōkōchi Masatoshi: Hito to sono jigyō* (Tokyo: Nikkan kōgyō shinbunsha, 1954), 11.

10. Ishibashi Tanzan, Ōkōchi Masatoshi, Katō Kōgorō, Matsui Haruo, and Komine Yanagita, "Nihon no shigen o kataru kai," *Kagakushugi kōgyō* (June 1939), 175.

11. Ibid., 155–156.

Kamei was interested not so much in Germany's specific technological inventions, but in the mindset and spirit that informed its inventions. He believed that ultimately the source of technological innovation was of a spiritual and political nature, not material and technical nature. The fount of scientific and technological innovation was the spirit of the ethnic folk (*Volk*) and its organic, national community (*Volksgemeinschaft*). Under Konoe's New Order, Kamei helped design a mass party to unite the Japanese people and mobilize the national spirit.

Japanese technocrats were convinced that Western-style empires, which had benefitted from their international comparative advantage in natural resources, were outdated. The new world trend was toward regional blocs dominated by countries that had achieved self-sufficiency in raw materials and commercial products through superior technology and organization. Technocrats argued that the new order would be based on different principles and practices. In contrast to western imperialism, which was based on the capitalist exploitation of colonies and semicolonies through monopolies, leaseholds, concessions, and "third party rights," the new ethnic-based regional orders would draw on managerial and spatial principles and strategies that aimed at the rational distribution, development, and overall profitability of the regional bloc. Among the organizational principles informing the new world order was the idea of spatially planning industry based on an assessment of the optimal location for production. In Manchuria, Japanese planners applied this concept in their policy of "suitable site for suitable industries," which aimed at the most rational division of industries within the Japan-Manchuria bloc. According to Kamei, Nazi regional planning was based on the multilateral organizational form (*takakuteki naru soshiki keitai*), in which industries within the bloc are managed in the same way as a technology-driven concern organizes its various subsidiaries based on a consideration of technology, management strategy, and factor costs.[12] Technocrats took up these concepts in the context of national land planning and the Greater East Asia Co-Prosperity Sphere.

With regard to mobilizing the national spirit, Japanese technocrats saw in European fascism a model of political organization that incorporated the latest managerial principles and mobilized the people. They contrasted the "organic" "leadership" organizations of German and Italian totalism with the "mechanical," "dictatorship" organization of communist (Soviet) totalism. According to one theorist the communist state was a "supra-racial, supra-national mechanical organization that took the individual as the core unit." In such a system, control was mechanical and impersonal because individual character was not

---

12. Ibid., 156.

recognized. The state resembled a feudal dictatorship in which a small elite and an unpopular leader oppressed the masses like slaves. In contrast, the fascist state was an organic racial state (*jinshu kokka*) based on ethnic self-determination. Control took the form of "living" control in which the part was not oppressed by the whole but rather its individual character was recognized as a vital, organic component of the whole. Whereas the unpopular leader Stalin relied on systems and institutions to "oppress" the masses, Hitler used his popularity and charisma to "lead" the masses. For this reason he described the Soviet Union as a "dictatorship state" or an "oppressed state." Germany, in contrast, was a "leadership state" (*shidō kokka*) or "managed state" (*keieisareta kokka*) which "manages with the utmost effort" in the same way that one manages a corporation.[13] Japanese technocrats envisioned Japan as a leadership or managed state in which "the state will be the manager and the state economy, as a whole, will be like one big corporation...the people are all employees."[14] With regard to the empire, they believed that "the preconditions for Japan to be the ruler (*meishu*) of the East were: a leader, centralization, and management with utmost effort (*kushin keieishugi*). We should deeply reflect upon the point that 'control is not dictatorship.'"[15]

Fascist Political-Economy

In contrast to military technocrats, who appealed to traditional Chinese principles of kingly way to promote the national defense state and economy, reform bureaucrats introduced the new theoretical approaches of political economy (*seijikeizaigaku*). These bureaucrats criticized classical economics as the so-called "economics of liberalism and individualism," which they claimed was misguided and based inappropriately on the notion of individual, rational decision-makers. They disparaged it as the "economics of exploitation," which served as a convenient "camouflage" of British colonialism under the theory of free trade and as the economics of the wealthy because it subordinated the country to the city under the name of commercialism and enabled finance capitalists to shift business risk to the masses through the mechanism of the stock market.[16] They were drawn to the study of political economy because it recognized the primacy of politics over the economy and attempted to find the most appropriate and

13. Katsuta Teiji, *Nihon zentaishugi keizai no seikaku* (Tokyo: Jitsugyō no Nihon), 13, 136, 144, 146.
14. Ibid., 136.
15. Ibid., 148.
16. Ibid., 182–183.

effective means to achieve state goals. Political economy revived the tradition of the German Historical School, especially its strong policy orientation and conception of the economy as an organically unified body. The historical school rejected the classical school's underlying theory of natural right, which viewed society and the state as composed of autonomous individuals. Instead it presented society as a living, developing organism in which the welfare of the whole takes precedence over the individual members.[17]

Reform bureaucrats were influenced by the writings of Werner Sombart and Friedrich von Gottl-Ottlilienfeld. From their prestigious posts at the University of Vienna and University of Berlin, respectively, these economists sought to provide the theoretical basis for the political and economic programs of the Nazi Party. Japanese bureaucrats found in their writings a way to ground political-economy within the ethnic national community and combined technology with ethnic nationalist spirit. From the mid-1930s to early 1940s, many of their works were translated into Japanese. The totalitarian ideas of Austrian economist and philosopher Othmar Spann also attracted considerable attention early on. By the late 1930s, however, his theory had fallen out of favor with the Nazis and was criticized by the Japanese for being too metaphysical and removed from actual life.[18] In wartime Japan, Sombart and Gottl were considered the leading representatives of the new economics with Gottl standing in a class by himself.[19]

Sombart gained a considerable following in Japan among planners and economists just as theoretical research on the Japanese managed economy was appearing. His works continued to be widely discussed and quoted by policymakers and economists well into the early 1940s.[20] Sombart was a student of Gustav Schmoller and associated with the "third generation" of the Historical School.[21] He began his professional career as a respected specialist of Marxism but later promoted his own brand of national socialism and anti-semitism under the name of "German socialism." His interests were wide-ranging but in connection to the reception of his ideas in Japan we can mention two aspects of his thought.

---

17. Horst Betz, "How Does the German Historical School Fit In?" *History of Political Economy* 20, no. 3 (fall 1998), 407–430.

18. On the German and Japanese reception of their ideas, see Erwin Wiskemann and Heinz Lütke, *Der Weg der deutschen Volkswirtschaftslehre: Ihre Schöpfer und Gestalten* (Berlin: Junker & Dünnhaupt Verlag, 1937); Kaneko Hiroshi, "Zentaishugi keizaigau no futakeikō—Gottoru to Shupan," *Kokumin keizai zasshi* 65, no. 2 (August 1938), 35–48.

19. Katsuta, *Nihon zentaishugkeizai no seikaku,* 179; Fukui Kōji, *Sei toshite no keizai* (Tokyo: Kōbundō shoten, 1937), 3.

20. On the reception of Sombart in Japan see Yanagisawa Osamu, "The Impact of German Economic Thought on Japanese Economists before World War II," in Yuichi Shionoya, ed., *The German Historical School: The Historical and Ethical Approach to Economics* (London: Routledge, 2001).

21. Betz, "From Schmoller to Sombart," 424.

First, Sombart put forth a schema of historical stages that could provide the basis for policymaking. Sombart offered a systematic approach to understanding the evolution of economic systems and emphasized the importance of historical and cultural influences. In his *Die Zukunft des Kapitalismus*, which was cited and discussed by Japanese planners, Sombart proclaimed that after World War I the age of "high capitalism" had ended and that the world economic system was in a state of transition. One possible form of the new economic system was a planned economy. His discussion of a planned economy that could accommodate both a market economy and a command economy was of great interest to Japanese planners.[22] In his *Deutscher Sozialismus*, Sombart attacked the previous liberal "Economic Age" as the "work of the devil" and criticized capitalist society for breaking up the natural organic community.[23] Sombart announced the coming of a new order based on the new national spirit of "German socialism." Second, like Spann, he viewed society as an organic entity, composed of estates with designated functions, but also argued that such a society required a strong state led by a forceful leader and managed by elites.[24]

In Japan, the most influential representative of the new national economy was Gottl. He remained an obscure figure in the United States and England and was heavily criticized by a number of contemporary German scholars.[25] Gottl aggressively attacked the teachings of the classical school of economics, and especially quantitative economics, which he felt was too abstract and static in its presentation and missed the dynamic relationships in society. Gottl advocated an economics based on actual life experience and the specific historical conditions and traditions of a country. He argued that the economy was not about property, value, or wealth, but about human communal life.

The central idea in Gottl's theory was his concept of *Gebilde*, translated as "socially constructed body" or "cooperative body" (*keiseitai, kōseitai, kyōdotai*). For Gottl, the eternal life form was human communal life, expressed originally

---

22. Yanagisawa, "The Impact of German Economic Thought," 175–178.

23. Werner Sombart, *A New Social Philosophy*, trans. of *Deutscher Sozialismus* (Princeton: Princeton University Press, 1937), 32.

24. See the article by Abram L. Harris, "Sombart and German (National) Socialism," *The Journal of Political Economy* 50, no. 6 (December 1942).

25. In Japan, he gained a following from the 1930s through works such as: *Wirtschaft als Leben; Der Mythos der Planwirtschaft;* and *Volk, Staat, Wirtschaft und Recht.* There were more than forty studies on Gottl written by Japanese between 1938 and 1941. In contrast to Spann and Sombart, he remains an obscure figure, although he has become the focus of several recent studies in German and Japanese, including Ursula Bender, *Technik: Technischer Fortschritt und sozioökonomische Zusammenhänge bei Friedrich von Gottl-Ottlilienfeld* (Frankfurt: Peter Lang Verlag, 1985); Ulrich Chiwitt, *Wirtschaft und Leben: Eine philosphische Analyse der Wirtschaftslehre Friedrich von Gottl-Ottlilienfeld* (Essen: Die Blaue Eule, 2000); Yoshida Kazuo, *Gottoru: Seikatsu toshite no keizai* (Tokyo: Dōbunkan, 2004).

in the household and state. The socially constructed body possessed an organic, dynamic relationship between itself and the individual members. In contrast to Spann, who conceived of the individual's relationship to the totality in static terms of part and whole, Gottl argued that the individual possessed his own unique world and socially unified body that was organically connected to the life of the larger socially constructed body. The life of the individual became a real possibility only via the cooperative body, which itself constructed social life only via the individual. Hence, the communal life inevitably demanded the subjective, internal union of every individual as a "fateful cooperative body" (*unmei kyōdōtai*).

According to Gottl, the three supreme socially constructed bodies were the ethnic cooperative body, the state, and the national economy. The ethnic cooperative body was both the natural and spontaneous socially constructed body and the fount of human communal life. Its existence and continuance was made possible on the one hand by the state, which created order out of the "life of ignorance" in human relations, and via the national economy, which created order out of the "life of poverty and distress" on the other. The ethnic cooperative body was foremost a politically unified state, based on the unified principles of freedom and control, and an economically unified national economy, founded upon the unity of demand and fulfillment (*Bedarf und Deckung*). Both the state and economy ultimately aimed at the preservation of the socially constructed body.

Gottl was a great admirer of Fordism and helped introduce its concepts to Germany. He believed that Ford's business philosophy of maintaining high wages and low-cost products served as the model of how firms should serve public principles. It represented a concrete example of how to "overcome the modern" by using ethnic nationalism to promote technology and vice versa. Gottl saw in Fordism a new principle for social change and the rise of a new kind of "leadership-type socialism" or "management socialism" based on the ideas of leaders and managers. He fiercely criticized the Soviet-type socialist system and its ideology of the planned economy as a delusion because it ignored the market, which Gottl saw as a legitimate, internalized regulating force that preserved economic order.[26]

Both Okumura Kiwao and Mōri Hideoto drew on the ideas of Gottl and Sombart to formulate their policies for Japan's New Order. These theories were useful to bureaucrats because they provided a theoretical basis from which to attack liberal capitalism, gave intellectual respectability to European fascism in Japan, and enabled them to present themselves as sophisticated, knowledgeable interpreters of the latest Western trends.

---

26. Yoshida, *Gottoru*, 31–39.

Okumura Kiwao

Okumura's training in German law and in the public service sector encouraged him to identify the state with social policy. He defined social policy as "the many kinds of institutions that improve the position and advance the welfare of society's lower class, not by destroying the foundations of the present economic and social structures, but by eliminating the insecurity of livelihood."[27] Okumura noted that "the basic idea of social policy is the repudiation of absolute freedom and uniform equality, while advocating both economic freedom and social equality."[28] His views of the state were similar to those of the Home Ministry's "social bureaucrats" in that he emphasized state social policy as a preventive measure; adopted a pragmatic, nationalist approach; and denounced left-wing socialism.[29]

Okumura's bureaucratic vision of social policy looked to the state as the primary vehicle to advance social welfare and national culture. He defended his ministry's proposal for a postal life insurance program as a means to assist the lower class. Okumura argued that in contrast to private life insurance, in which profit was the first principle, state life insurance accepted all applicants regardless of situation and used the funds to improve social welfare. Moreover, state life insurance served as both a form of compulsory savings for the poor and a vehicle to channel funds toward revitalizing the regions.[30] Okumura also took up the cause of workers in the public service sector. He congratulated the Home Ministry's Social Bureau on its progressive labor policies. In the communications sector, Okumura called for better treatment of workers, a minimum wage, and labor unions. He justified these policies based on the conviction that "today, the state's goals are not 'power' and 'law,' but 'cultural goals' and 'welfare.' In order to protect society's lower classes, the state should go beyond the conflicting interests of individuals and appropriately intervene in the private economy."[31]

Okumura argued that the state, not private industry, should manage the communications sector. Based on considerations of efficiency and that sector's vital role as "the harbinger of culture and the eyes and ears of society," public interests were better served by the state. As he explained, in the case of postal service, "the delivery of mail must not only be prompt, accurate, respond widely to the

---

27. Okumura Kiwao, "Shakai seisaku ni okeru kani hoken no shokunō," in Okumura Kiwao, *Teishin ronsō* (Tokyo: Kōtsū kenkyūsha, 1935), 400–401.

28. Ibid., 400.

29. See Kenneth Pyle, "The Advantages of Followership: German Economics and Japanese Bureaucrats, 1890–1925," *Journal of Japanese Studies* 1, no. 1 (autumn 1974), 127–164.

30. Okumura, "Shakai seisaku ni okeru kani hoken no shokunō," 395–406.

31. Okumura Kiwao, "Teishin gengyō koyōin taigū kaizen no ikkōsatsu," in Okumura, *Teishin ronsō*, 392.

demands of the public, and be popularized throughout the country, but the fees must also be low." State management was most appropriate because of the need for speed, accuracy, broad application, and low fees, as well as to protect privacy, coordinate international delivery, and utilize public transportation. Okumura believed that private companies seeking to obtain profits, lacked sufficient incentives to keep fees low. Given their monopoly over postal services, they would probably raise fees. Moreover, private companies would not foster the government's mission of broadening access to postal service throughout the country and would probably concentrate their services in the most profitable urban areas at the expense of the regions.[32] Okumura made the following case for public over private management:

> The activities of the modern state are not limited to the area of political affairs in which it seeks to exercise its sole sovereignty, but also include the broad public sector which aims at the people's welfare. The reason for the government's management of the communications industry lies here. In other words, from a country's political standpoint, communications organs were from the beginning indispensable cultural, economic, and industrial organs; their role in assisting the ruler uphold good and suppress evil and participating in the people's welfare have been extremely great. Considering the mission of communications organs, they cannot be entrusted to private management based on "profitism" (*eirishugi*) and should be managed by the state itself based on public interest (*kōekishugi*).[33]

During the early 1930s, Okumura's views of the social role of the state meshed well with the army's demands for a national defense state. Okumura developed close ties with Suzuki Teiichi, who also sympathized with the lower classes. Suzuki, then head of the Manchuria-Mongolia Affairs Unit, requested Okumura's assistance in establishing Manchurian Telegraph and Telephone.[34] Okumura supported MTT's mission to "contribute to the great project of establishing Manchukuo and advance the welfare of the Japanese and Manchurian people" and "bear one wing of broad national defense."[35] He affirmed the army's view of the public character of the special company and the desire to prevent zaibatsu

32. Okumura Kiwao, *Yūbin hō ron* (Tokyo: Katsumeidō shoten, 1927), 1–4.

33. Okumura Kiwao, "Tsūshin tokubetsu kaikei setsuoku no rironteki konkyo," in Okumura, *Teishin ronsō,* 189–190.

34. Okumura Katsuko, *Tsuioku Okumura Kiwao* (Tokyo: Okumura Katsuko, 1970), 42.

35. Manshūseifu ed. *Manshū kenkoku jūnenshi,* 903; Okumura, "Manshū denshin denwa kabushikigaisha no setsuritsu," in Okumura, *Teishin ronsō,* 142.

domination. He emphasized that "MTT is not simply a profit-based company but a national policy organ."[36]

As chief of the Wireless Division at his ministry's Electrical Affairs Bureau from June 1934, Okumura promoted the use of radio broadcasting as a key propaganda tool to shape foreign opinion on Japan.[37] Japan's diplomatic relations had reached a nadir after its withdrawal from the League of Nations over the issue of Manchuria. Okumura called for the creation of an information and propaganda organ in order to effectively utilize new mediums like wireless communication for propaganda.[38] He proposed a merger of Japan's two major wire services, Rengō and Dentsu, in order to present a unified front toward the outside world, as well as the establishment of a system for censoring news content and strengthening of censorship organizations in Japan. In November 1935, the government created the state-run United News Agency. The financially weak Rengō readily merged its operations into the new agency. The more profitable Dentsū initially resisted, but eventually capitulated in the face of the government's strong-arm tactics.[39] Consolidating state control over radio was a relatively easy affair, since the state had dominated radio from the beginning. Securing state control over its energy source, electric power, was an entirely different matter.

The Electric Power Control Law

When Okumura presented his draft for state control of electric power in 1935, based on the principle of "private ownership, public management" (*minyū kokuei*), business accused him of promoting "national socialism" and "bureaucratic fascism." Okumura did eventually identify the idea of separating ownership and management with fascism, but initially, he justified it on the basis of public interest. Okumura's draft was the product of collaboration with other technocrats at the Cabinet Research Bureau. Originally conceived as the administrative wing to the liberal Cabinet Deliberation Council under Prime Minister Okada, the bureau acquired the function of the military's economic general staff. It was expanded into the Cabinet Planning Agency and later the Cabinet Planning Board. Its members included a number of prominent left-wing technocrats

---

36. Okumura Kiwao, "Manshū denshin denwa kabushikigaisha no denpō ryōkin," in Okumura, *Teishin ronsō*, 162–163.

37. Okumura Kiwao, "Nichi-Man kan no tsūshin kankei," in Okumura, *Teishin ronsō*, 83.

38. Okumura Kiwao, "Musen hōsō to kokusai jōhō," in Okumura, *Teishin ronsō*, 39.

39. Gregory J. Kasza, *The State and the Mass Media in Japan, 1918–1945* (Berkeley: University of California Press, 1988), 153–157.

such as Suzuki Teiichi, Katsumata Seiichi, Takahashi Kamekichi, and Kawakami Jōtarō.[40]

The importance of electric power for national defense became apparent to Okumura in his efforts to bolster the state's propaganda campaigns. In order to broadcast radio programs overseas, Japan required large electric power stations. Okumura reported that foreign countries had been rushing to establish large electric power stations in order to dominate domestic radio waves, obstruct incoming foreign propaganda, and transmit their own programs abroad.[41] The Soviet Union was devoting considerable efforts in this area by building the world's largest 500-kilowatt electric power transmitter in Moscow, in addition to seven 100 kilowatt stations in various cities. He noted that plans were currently underway in Japan to revise its small-scale, decentralized system and build a 150-kilowatt station in Tokyo.[42] Okumura argued that it was crucial to place the industry under state control on the grounds that electricity was the decisive force behind a flourishing munitions industry and the basic resource for the chemical industry, which was essential from the military standpoint.[43]

From a managerial standpoint, state control over electric power represented the wave of the future. The increased demand for electric power during the 1920s and 1930s helped launch an administrative revolution in the industrial world. New programs for state management of electric power emerged in response to the technical requirements for efficient power generation and distribution. England's "grid system," America's Tennessee Valley Authority (TVA), and the Soviet Union's first Five Year Plan for electric power generation reflected new managerial visions of social and regional planning, state regulation, and integrated resource development. Behind these programs was the assumption that modern society had entered the era of planning and control and that the public services sector, in particular, was incompatible with free competition.[44]

Okumura proposed that the generation and transmission of electric power be centralized and standardized to provide cheap and abundant electric power for civilian and military use. At the time, there were more than 650 small electric companies using different technology and charging different rates. The five major utility firms were heavily indebted, partly resulting from the cutthroat competition in accessing water rights and expanding service areas, both secured

40. Itō Takashi, "'Kyokkō itchi' naikakuki no seikai saihensei mondai," 60–61.

41. Okumura, "Musen hōsō to kokusai jōhō," 45.

42. Ibid., 46.

43. Okumura Kiwao, *Denryoku kokuei* (Tokyo: Kokusaku kenkyūkai, 1936), 3–4.

44. See Philip Selznick, *TVA and the Grass Roots: A Study in the Sociology of Formal Organization* (New York: Harper Torchbooks, 1966); Melvin G. de Chazeau, "Rationalization in Electricity Supply in Great Britain," *Journal of Land and Public Utility Economics* 10, no. 3 (August 1934), 254.

through patronage of local party officials. Okumura argued that the state was best equipped to coordinate the simultaneous processes of generation and transmission of electricity and to build and manage a national system of high-power transmission lines. In the case of hydroelectric power generation, which was a key energy source, the state had the resources to build large hydroelectric power plants to make use of Japan's abundant water supplies.[45] Okumura was attracted to the British model of separating generation and transmission of electric power from distribution. Its government managed the network of gridlike primary transmission lines via a nonprofit, publicly owned corporation, while leaving the distribution side to private business. Such a system ensured that similar considerations of efficiency and public welfare—that is providing abundant electricity at low rates—took precedence over maximizing profits, while leaving the system of private property intact. In addition, the United States' ambitious TVA project to develop the water resources of the Tennessee River offered a useful planning model for regional resource development in Japan as well as Manchuria. It suggested a new type of administrative program combining public management to guarantee public interest and order, and private ownership to preserve incentive and flexibility.[46] Okumura rejected the Soviet model of state ownership of electric power facilities because he wanted to avoid increasing public debt, administrative personnel, and bureaucratic red tape.[47] Okumura also took pains to distinguish his plan from Soviet-style socialism.

Okumura's concept of "private ownership and public manpower" was influenced by the famous 1933 study of American New Dealers Adolf Berle and Gardiner Means.[48] According to Berle and Means, the two hundred largest corporations in the United States controlled the largest share of corporate and national wealth and had become quasi-public institutions. The owners, represented by hundreds and thousands of stockholders, were removed from the day-to-day operations of the firm. Managers, who no longer had a direct claim to the majority of profits, lacked the incentive to manage efficiently and maximize profits or, at worst, tried to enrich themselves at the expense of shareholders. The traditional relationship between property ownership, profits, and performance no longer

---

45. Okumura Kiwao, "Denryoku kokukan mondai," in Andō Yoshio, ed., *Shōwashi e no shōgen*, vol. 3 (Tokyo: Hara shoten, 1993), 152.

46. Selznick, *TVA and the Grass Roots;* Norman Wengert, "TVA—Symbol and Reality," *Journal of Politics* 13, no. 3 (August 1951), 269–392.

47. Okumura Kiwao, "Denryoku tōsei no saishuppatsu" (November 1940), in Okumura, Kiwao ed., *Henkakuki Nihon seiji keizai* (Tokyo: Sasaki shobō, 1940), 280.

48. Ibid, 279; Adolf A. Berle and Gardiner C. Means, *The Modern Corporation and Private Property* (New York: Macmillan, 1933). See also Robert Hessen, "The Modern Corporation and Private Property: A Reappraisal," *Journal of Law and Economics* 26, no. 2 (June 1983), 275–278; Ellis W. Hawley, *The New Deal and the Problem of Monopoly* (Princeton: Princeton University Press, 1966), 173.

existed. In the modern economy the large corporations were no longer efficient players in a free market and the profit motive no longer promoted society's well-being. Moreover, the concentration of economic power among a small group of corporate directors and executive officers within the largest corporations and their monopolistic behavior proved that free competition was a fiction. Their study concluded that a new logic of the quasi-public corporation was emerging in which property ownership was secondary and separate from management. Public interests of technical efficiency and social responsibility, not private profits, should be the main incentive and the state must take a greater role to ensure that public interests were advanced. For Japanese technocrats, the principle of the "separation of ownership and management" opened up a third way between liberal capitalism and socialism.

Not surprising, the two leading electric companies were less enamored of this approach. The presidents of Tokyo Electric and Fuji Electric led the counterattack against the reformist plan, backed by the other major electric firms and business associations such as the Industrial Club, Japan Economic Federation, and Japan Chamber of Commerce and Industry.[49] The business world drew on its political connections to Commerce Minister Ogawa Gōtarō and Railway Minister Maeda Yonezō, who participated in the Four Ministers Conference to deliberate on the plan. Ogawa had close ties to the Minseitō and Osaka business and Maeda was an influential Seiyūkai leader, who sat on the board of a number of electric companies.[50] The navy adopted a cautious stance toward the plan, while the ministers of agriculture and education rejected it outright. The few supporters of the plan were the army, the Communications Ministry, the Social Masses Party, and the new zaibatsu. The communications bureaucrats wanted to increase their administrative control over electricity and water rights. Katayama Tetsu of the Social Masses Party hailed it as a constructive step toward socialism. The new zaibatsu welcomed cheaper electric power for its industries.[51]

The plan was denounced not for its economic arguments, which made good sense, but for its association with the military's fascist blueprints for reform.[52] Shortly after the plan was presented, radical officers staged the February 26 incident. Moreover, as the first major plan put forth by the new Hirota cabinet, which had come to power on the platform of "comprehensive government

49. Ogimura Ryūtarō, "Denryokuan hantai no nami," *Keizai ōrai* (October 1936), 316–317.

50. Okumura Kiwao, "Denryoku mondai kaiketsu no kagi," Interview, in Andō Yoshio, ed., *Shōwashi no shōgen, vol. 3* (Tokyo: Hara shoten, 1993), 160.

51. Katayama Tetsu, "Denryoku kokuei no mokuhyō to tsūshinsho an," *Nihon hyōron* (Special edition: October 1936), 387–392; Ayukawa Yoshisuke, "Yo no tōsei keizai kan," *Jitsugyō no Nihon* (August 1938), 18–22.

52. Okumura Katsuko, *Tsuioku Okumura Kiwao*, 42–49.

reform" (*shōsei isshin*), it was seen as a harbinger of future government reform. The plan established political precedents in terms of how policies would be made and whose interests would be represented. It went farther than previous plans in asserting the primacy of the state's interests and permitting it to directly intervene in the internal affairs of private enterprise. It symbolized the battle line between supporters of the "status quo" and private interests and advocates of "reform" and public interests. Okumura admitted, "At present, whether state management of electric power will be implemented or not will determine the political direction taken at the crossroads of maintaining or rejecting the status quo, totalism or individualism, and controlled economy or laissez-faire."[53] He described the electric power controversy as a major "thought war" taking place within the nation.[54]

Okumura was shaken by the vehement opposition of the main parties and big business to his plan and the personal attacks on himself and fellow planners. The cabinet failed to win Diet approval despite the determination of the Hirota cabinet to make it a top policy objective. Attempts to push the bill, together with the army's administrative reform proposal in the seventieth Diet session of December 26, 1936, ended in the bitter exchange between the Seiyūkai leader Hamada Kunimatsu and Army Minister Terauchi, followed shortly by the fall of the Hirota cabinet and replacement by the short-lived Hayashi cabinet. Not until the formation of the Konoe government in April 1937 were constructive steps taken to pass the law. Important concessions were made to business, including increasing business oversight through the new advisory council, financial guarantees by the state for dividends and stocks of the new entities, and exclusion of auto generation and transmission.[55] The new law was a bittersweet victory for technocrats. They succeeded in passing the first state control law, but also got a taste of the battles ahead. The lesson that Okumura drew was that strong leadership and some form of direct popular mobilization was needed to win acceptance for state control.[56]

The Totalist State

Okumura embraced more radical and authoritarian conceptions of state and society after observing European fascism firsthand in 1937. In a broadcast shortly

53. Okumura Kiwao, "Genkō denki gyōsei no kekkan to kokuei mondai," in Okumura Kiwao, ed. *Denryoku kokuei* (Tokyo: Kokusaku kenkyūkai, 1936), 62.

54. Ibid., 161–162.

55. Richard J. Samuels, *The Business of the Japanese State* (Ithaca: Cornell University Press, 1987), 147.

56. Okumura, "Denryoku kokkan mondai," 161.

after his return, Okumura reported that of all the countries he visited, he was most impressed with Nazi Germany and Fascist Italy because of their enormous efforts and apparent successes in rallying their nations behind state goals. In Nazi Germany, the concept of "primacy of public over private interests" (*Gemeinnutz geht vor Eigennutz*) was raised to a first principle.[57] Okumura praised the Mussolini government's ability to mobilize patriotic sentiment by appealing to the spirit of ancient Rome.[58] He was struck by the total commitment of the German people to building a national defense state. As he noted, with the spirit of national sacrifice, expressed via the national slogan "guns before butter" and the "movement to eliminate waste," the people were reducing their consumption of imported basic food products in order to save precious gold reserves and were devoting their energies to munitions production.[59]

More than their specific policies and programs, Okumura was drawn to their ideology, which he variously referred to as the "totalist state view" (*zentaiteki kokkakan, zentai kokka no shisō, zentai kokkashugi*), "totalism" (*zentaishugi*), and "fascism."[60] According to Okumura, this ideology

> does not view the state as merely an assembly of individuals, but as the highest virtue. Furthermore, it establishes a control system that rejects laissez-faire and emphasizes the spirit of public welfare to serve the state and society by abandoning personal profit-oriented egoism, and rallies the cooperative spirit of the people by eliminating the notion of class conflict by replacing individualism with totalitarianism.[61]

He now identified managerial concepts with fascism:

> I am convinced that from now on the spirit of the civilization and politics of mankind is fascist ideology. The essence of fascist thought is the worldview which exalts public interest based on the totalist state view. In the economic sphere, it is an economy based on so-called "controlism" as opposed to "liberalism." I think that the trend of the controlled economy is toward "private ownership, public management" (*minyū kokuei*).

57. On the reception of the concept "Gemmeinnutz geht vor Eigennutz" in Japan see Yanagisawa Osamu, "'Gemeinnutz geht vor Eigennutz'" im Streit um die Neue Wirtschaftsordnung in Japan in der kritischen Zeit," in Rainer Gömmel and Markus A. Denzel, eds., *Weltwirtschaft und Wirtschaftsordnung: Festschrift für Jürgen Schneider zum 65 Geburtstag* (Stuttgart: Franz Steiner Verlag, 2002), 301–314.

58. Okumura Kiwao, "Itarī to Doitsu no inshō," in Okumura, *Henkakuki Nihon seiji keizai,* 40–41.

59. Ibid., 46–53.

60. Okumura Kiwao, "Ō-Bei kansatsu yori kaerite," in ibid., 56; Okumura Kiwao, "Tenkanki ni tatsu sekai to Nihon," in ibid., 90.

61. Okumura, "Itarī to Doitsu no inshō," 53; Okumura, "Ō-Bei kansatsu yori kaerite," in ibid., 56.

This was my conviction before going [abroad], and these feelings have become increasingly stronger since.[62]

Okumura announced that the world was at a historical turning point.[63] He likened this great change to a "world-historical cultural revolution" characterized by "a revolution in law, politics, society, and economy." According to Okumura, "before the iron laws of historical development, the downfall of the liberalistic, individualistic, capitalistic world is unavoidable. A new, higher cultural system is replacing or trying to replace the existing individualistic system."[64] Quoting Sombart, he described the present mass society as one characterized by the concentration of capital and atomization of humans. He likened the big cities and industrial belts to sand dunes, in which humans gather together without mutual bonds like grains of sand. For Okumura, this atomization of humans both at the national and international level was the fundamental characteristic of the individualistic economy and the "tragedy of the twentieth century."[65] Okumura looked to a new order, which he described as one that "will probably possess a different shape—economically, culturally, and geographically—from the present world." Under the new order:

> The reconstructed social structure will be a new, more rational, organizational, and powerful basic structure that goes beyond the isolation, desolation, and fragmentation of the present individualistic economic organization. The productive power achieved under capitalism will be rationally applied for the purpose of the state and society, and will be rapidly expanded in line with state goals.[66]

Okumura advanced a techno-fascist program for an advanced national defense state. He advocated planning and organization on a national scale and national unification under a totalist order.[67] In terms of the economy, he argued that "the course that Japan should necessarily pursue from now on is one along the lines of a totalist planned economy. A rapid reorganization of the economic and state structure must be effected. Across the board strengthening and streamlining of the entire structure is of urgent necessity."[68] Okumura emphasized that both spiritual and material mobilization were needed in order to create a

62. Ibid., 55.
63. Okumura, *Nihon seiji no kakushin*, 3–4.
64. Ibid.
65. Ibid., 5.
66. Ibid., 8–9.
67. Ibid., 99.
68. Okumura, "Tōa kensetsu hyakunen sensō," 147.

totalistic planned economy. Like Nazi Germany and Fascist Italy, Japan needed to unite ideologically and replace the profit-oriented ideology (*eirishugi shisō*) with an ideology that places priority on public interests (*kōekishugi shisō*). Once the economic substructure was transformed into a totalist planned economy, the ideological superstructure would also change, giving rise to a totalist thought system.[69] Okumura was vague about the specific type of economic system to be created.

As for the government, Okumura called for a strong reformist cabinet that could push through controversial reforms and obtain the trust and cooperation of the people. Such a cabinet must achieve comprehensive control over all aspects of the state, provide active leadership, respond promptly to the changing circumstances, and obtain popular support. By active leadership he meant restricting freedom of speech and discussion, fostering a unified political ideology or "unified nationalism," controlling trade, and raising taxes.[70] In order to strengthen executive leadership, he recommended that the cabinet be streamlined and centralized by reducing the number of ministers of state and strictly separating their role from that of administrative heads. Okumura also demanded a new system to train and promote bureaucrats and the replacement of generalist bureaucrats with technical specialists. In hinting at the plans for a new Political Order, he proposed to transform the Diet into an "imperial works assistance organization" and create a new party that would incorporate the industrial unions, youth groups, and agrarian reform groups and absorb the large number of soldiers returning from battle.

Mōri Hideoto

More than any other ideologue of wartime Japan, Mōri articulated a vision that fused left-wing visions of labor, right-wing pan-Asianist and geopolitical views, and technocratic leadership. Mōri's eclectic approach reflected his diverse background as social activist, finance bureaucrat, reformist ideologue, and economic advisor to the army. He cultivated a broad network of bureaucrats, army officers, right-wing activists, engineers, progressive intellectuals, journalists, and labor party leaders. His residence in Kamakura became a popular meeting place for reformists of various political hues. Mōri described himself and his colleagues as "economic technologists" who were bringing "economic technology" to Asia and, in the process, were creating a new type of society. He claimed that

---

69. Okumura, *Nihon seiji no kakushin*, 123.

70. Ibid., 128–141.

Japan would "liberate" Asia from Western capitalist imperialism by replacing the old class-based society organized on the basis of capitalists versus workers, or owners versus producers, with an organic, hierarchical, functionalist society in which managers and laborers were all organizers. Class consciousness would be replaced by ethnic consciousness, in which every member was a vital organ of society and had a function and status. Moreover, the ethnic nationalist spirit of unity would serve as the fount of scientific creativity and technological innovation that was the key to Asia's independence from the West.

Mōri's techno-fascist vision was a product of three intellectual phases in his life. Each phase was associated with a particular person or group, place, event, and institution and added a new dimension to his thought. Viewed together, these overlapping phases mark his ideological shift to the right and successive stages in the formation of his fascist vision. In his socialist phase, Mōri developed an anticapitalist perspective at Tokyo University through his political activism and ties to right-wing labor activists, especially Kamei Kan'ichirō. His pan-Asianist phase began in 1933, when Mōri was posted to Manchuria and then North China, where he embraced the pan-Asianist and geopolitical views of Kwantung Army officers. Mōri's technological phase began after he returned to Japan in 1938. He read deeply and wrote on Japan's technological mission in Asia and worked closely with Japanese government engineers to design Japan's New Order for Science-Technology.[71]

Mōri's socialist phase began as a student of law and politics at Tokyo Imperial University during the heyday of Taishō liberalism. He participated in the Tokyo Imperial University Settlement project, one of the major projects of the left-wing student group, Shinjinkai.[72] The Settlement was established for the purpose of improving the conditions of the poor in the Yanagishima district in Honjo ward. It provided various community services such as medical assistance, education facilities, and a cooperative living experience for workers and students. Shinjinkai members also established a labor school to instill working-class consciousness among the poor. Such experiences shaped his anticapitalist, populist social consciousness.

Through his political activism, Mōri became acquainted with technocratic intellectuals and labor leaders, most notably Kamei Kan'ichirō. A former diplomat and founder of the Social Masses Party, Kamei acted as Mōri's mentor and promoted his career. He claimed to have arranged for Mōri to enter the prestigious Finance Ministry, where he worked in tax administration. In 1933 Kamei used

---

71. Suzuki Teiichi, "Suzuki Teiichi shi danwa sokkiroku," 351–352.
72. Henry Smith, *Japan's First Student Radicals* (Cambridge, MA: Harvard University Press, 1972), 142–44.

his connections to Hoshino Naoki to arrange for Mōri's transfer to Manchukuo.[73] Kamei described Mōri as his "confidante" and "brain."[74] After returning to Japan in 1938, Mōri joined Kamei's political study group, which was formed at the behest of Konoe to explore the possibility of creating a mass party.

In Manchuria and China, Mōri served as a political strategist, planner, and a liaison between bureaucrats and the military. In Manchukuo, Mōri held joint posts at the General Affairs Bureau as head of special accounts and at the Financial Department as head of tax policy. He was unusually popular with army leaders, who invited him to serve as the economic advisor to Japanese military headquarters in Tientsin in May 1937 and to the army's Special Affairs Section in Beijing at the time of the China Incident. Mōri also formed close ties to the radical right.[75] In 1935 he became acquainted with the right-wing journalist Sugihara Masami, who founded the reformist journal *Kaibō jidai* (*Era of Analysis*). After returning to Japan, Mōri developed his fascist vision in regular contributions to this journal under the penname Kamakura Ichirō.[76]

Mōri's extended service in Manchuria and China led to a strong identification with the pan-Asianism and geopolitical views of army officers. Shortly after his arrival in Manchuria, Mōri reportedly claimed that Manchuria had awakened his intellect and spirit, especially his "respect and affection" (*keiai*) for the Chinese, Mongolian, and Korean people in Manchuria.[77] However, recent archival findings have provided a very different image of Mōri, who, along with many other Manchukuo leaders such as Tōjō Hideki, Kishi, and Furumi Tadayuki, has been connected to Manchukuo's lucrative opium monopoly trade run by the underworld businessman Satomi Hajime.[78] More than other bureaucrats and officers who served in Manchuria and China, Mōri was seen as a pan-Asianist. He consciously cultivated this image by wearing Chinese clothes outside of work and pacing the halls of the Finance Ministry in contemplative thought.[79] Suzuki Teiichi described Mōri and Colonel Mutō Akira, then chief of military affairs, as intellectual opposites. He associated Mōri's thought with that of the

73. Hoshino, *Mihatenu yume*; Kamei Kan'ichirō, "Kamei Kan'ichirō danwa sokkiroku," 37.

74. Ibid., 35–39.

75. Kan, Interview, 247.

76. Itō Takashi, "Mōri Hideoto ron oboegaki," reprinted in *Itō Takashi, Shōwaki no seiji (zoku)* (Tokyo: Yamakawa shuppankai, 1993), 130.

77. Mōri Hideoto, "Tairiku no kanki: Shikaisaru ni shinobinai—watakushidomo no shinjitsu na kimochi," Osaka mainichi shinbun, Document 234, *Mōri Hideoto kankei bunsho*.

78. Yamamuro, *Manchuria under Japanese Domination*, 231, 266; Reiji Yoshida, "Japan Profited As Opium Dealer in Wartime Japan," *Japan Times* (August 30, 2007).

79. Hata Ikuhiko, *Kanryō no kenkyū* (Tokyo: Kōdansha 1983), 130–131; Nihon hyōron shinsha, ed. *Yōyōtaru*, 175.

"East Asian cooperative body," while describing Mutō as a devoted—and rather uncritical—follower of Nazism.[80]

Mōri's technological views were shaped by a variety of experiences, especially his development and intelligence work in Asia, readings of Gottl, and collaboration with engineers at the Asia Development Board. His vision was influenced by Kamei, who became a leading proponent of German geopolitics and the Nazi technological worldview especially during the late 1930s.[81] At the Asia Development Board, Mōri was appointed first section chief of the Economic Division and worked closely with Miyamoto Takenosuke. Together they drafted plans for the New Order for Science-Technology. During the late 1930s, Mōri wrote extensively on the China Incident and Japan's technological mission and geopolitical destiny in Asia. He was influenced by the writings of Gottl and sought to apply his concept of *Gebilde* to Japan's New Economic Order and the East Asia cooperative body. Through his writings, speeches, and interviews on Japanese technology and East Asia, he engaged in a dialogue on technology with prominent left-wing intellectuals such as Miki Kiyoshi, Aikawa Haruki, and Tosaka Jun.

The East Asian Cooperative Body

Around the time of Mōri's return to Japan in the spring of 1938, military and civilian technocrats in Japan were repositioning themselves on the battle in China to address the changed circumstances. After nearly a year of fighting, Japanese troops had failed to force the capitulation of Chiang Kai-Shek. It had become clear that military might alone was not enough to win the war, especially given Japan's limited supplies of materiel and foreign exchange reserves. Due to the stiff resistance by the Chinese and increasingly sharp criticism from abroad, especially from the United States, Japan now sought to play a direct role in "state-building" in China by establishing a pro-Japanese government and a new ideology that could promote Asian nationalism and serve as a basis for propaganda.

On the home front, a more effective vehicle than Konoe's heavily bureaucratic National Spiritual Mobilization Movement (Kokumin seishin sōdōin undō) was needed to sustain popular morale.[82] In the spring of 1938, frustrated by the fierce opposition exhibited in the grueling seventy-third Diet session, Konoe requested a group composed of Kamei Kan'ichirō, Asō Hisashi, Akiyama Teisuke,

---

80. Suzuki, ed., "Suzuki Teiichi shi danwa sokkiroku," 352–353. Mutō was an active figure in the Imperial Rule Assistance Association in Japan in the early 1940s.

81. Kamei claimed to have studied under Karl Haushofer during his diplomatic posting in Germany in the 1920s.

82. The National Spiritual Mobilization Movement, founded in the fall of 1937, was administered by the Cabinet Information Bureau, Home Ministry, and Education Ministry.

and Akita Kiyoshi to draw up plans for a mass party. In the summer of 1938 the Shōwa Research Association launched the Cultural Problems Research Group under Miki Kiyoshi to establish a new ideological principle for China policy.[83] The recently established Cabinet Planning Board immediately began drafting plans for a new planning organization for China affairs to provide unified civilian and military leadership over the political, economic, and cultural affairs of the expanding occupied territories in China.[84]

From the fall of 1938 onward, Mōri set out his vision of an East Asian cooperative body in essays published in *Kaibō jidai*. Mōri and Sugihara shared similar views on China and together they defined the journal's basic stance toward the war. According to Sugihara, he and Mōri first came up with the idea of an East Asian cooperative body sometime after their first meeting in Manchuria.[85] A month after returning to Japan, Mōri gave a talk on the subject at Yatsugi's National Policy Research Association entitled "Towards the Development of an Eastern Cooperative Body."[86] Members of the Shōwa Research Association were also promoting the idea of a "cooperative body" in Asia.[87]

Mōri's vision of the East Asian cooperative body appealed to the notion of a common Asian heritage and the cooperation or community of states within a unified organic body. Like Okumura, he described the war as Asia's revolution against the West and "a holy war for the purpose of eternal peace in the Orient."[88] For Mōri, the war's historical significance lay in its attempt to "overcome the modern."[89] He argued that Japan was seeking to build a new order that would be based neither on capitalism nor communism, but on ethnic nationalism. This new order represented a "third empire" or "third order," which he described as "a comprehensive living order of the Japanese and Chinese ethnic peoples."[90]

83. Fletcher, *Search for a New Order;* Itō Takashi, *Shōwa jūnendaishi danshō* (Tokyo: Tokyo daigaku shuppankai, 1981), 21–29.

84. Baba Akira, *Nitchū kankei to gaisei kikō* (Tokyo: Hara shobō, 1983), 356; see the introduction by Imura Tetsuo, in Imura Tetsuo, ed., *Kōain kankō tosho-zasshi mokuroku: Jūgonen sensō jūyō bunken shirizu,* no. 17 (Tokyo: Fuji shuppan, 1994).

85. Itō Takashi, "Shōwa jūsannen Konoe shintō mondai kenkyū oboegaki," in Nihon seiji gakkai, ed., *"Konoe shintaisei" no kenkyū* (Tokyo: Iwanami shoten, 1972), 163.

86. The talk given in June 1938 was entitled, "Toward the Development of an Oriental Cooperative Body." See Yatsugi Kazuo, *Shōwa dōran shishi,* vol. 1 (Tokyo: Keizai ōraisha, 1978), 490–493.

87. Fletcher, *Search for a New Order,* 110–116.

88. Kamakura Ichirō, "'Tōa kyōseitai kensetsu no shojōken: Chōki kensetsu no mokuhyō," *Kaibō jidai* (October 1938), 24.

89. Sugihara Masami, "Ni-Shi jiken ni kataserareta sekaishiteki kadai," *Kaibō jidai* (July 1938), 9.

90. Ibid., 29, 30.

Mōri saw ethnic nationalism as an active, progressive force."[91] For the Japanese, he wrote, the war represented the "self-awakening of ethnic consciousness of the Japanese ethnic peoples" and the "lively act of discovery and manifestation of the 'ethnic self.'"[92] As Mōri explained, "the life development of the Japanese ethnic peoples, while being completed within the scope of each historical stage, had finally reached the historical stage via the China Incident in which it should be completed on a world-scale."[93]

> The development of the Japanese ethnic peoples is the extension of the act of eternally continuing to live a new life while preserving something completely intrinsic or ancient of the ethnic peoples that exists within its essence. This brings to mind the life force of the two-sided Janus, which eternally possesses a youthful appearance. The life force of the Japanese ethnic peoples, however, is not that of Janus, possessing two separate life forces, but a unified life force in which something intrinsic and inherent is made to act eternally toward a new, youthful life. For this reason, the act of vital, rapid advance of the Japanese ethnic people is concurrently expressed by the impulse to develop a new life and to try to preserve that original Japan.[94]

By suggesting that the Japanese ethnic peoples could retain their essence and constantly develop, Mōri appealed to the right-wing call to preserve the *kokutai* and technocratic concerns to secure a self-sufficient sphere in East Asia.[95]

Mōri argued that an ethnically based New Order in East Asia required strong, autonomous states that could stand up to the West. He set forth two basic conditions for the establishment of an East Asian cooperative body: the political unification of China and the reform of Japan itself. With regard to the first condition, he took pains to differentiate Japan's policy toward China from the so-called "imperialism" of the West. Mōri admitted that previously Japan had practiced Western-style imperialism in China. However, he argued, as a result of the Manchuria Incident, Japan's relations to north China began to take on a different character—not a diplomatic, commercial relationship as in the past but one "oriented toward the masses."[96] According to Mōri, before the outbreak of

---

91. Kamei, "Doitsu no sangyō seishin," 158.
92. Kamakura, "'Tōa kyōseitai kensetsu no shojōken," 24.
93. Kamakura Ichirō, "Kokumin soshiki to Tōa kyōdōtai no fukabunsei," *Kaibō jidai* (January 1939), 22.
94. Ibid., 22; see also Mōri Hideoto, "Shina no sangyō kaihatsu," *Tōyō* (August 1939), 80–81.
95. See the analysis of Mōri's conception of ethnicity in Itō, "Mōri Hideoto ron oboegaki," 131–132.
96. Kamakura, "'Tōa kyōseitai kensetsu no shojōken," 26.

the China Incident, Japan's foreign policy was bifurcated. Its relationship toward north China represented the policy of "continental Japan" (*tairiku Nihon*), by which Japan concerned itself with the "political character" of China by founding Manchukuo.[97] In central and south China, however, Japan continued the Western imperialist policy of "maritime Japan" (*kaiyō Nihon*) by which it expanded and maintained its liberal capitalistic order.[98]

Mōri asserted that the true objective of the current war was to pursue the new policy of continental Japan. For this reason, the new planning organization for China affairs was named the Asia Development Board.[99] Its founder, Suzuki Teiichi, had also explained that the aim was the social and economic stability of China; "Asia development" (*kōa*) meant "making a complicated, colonial China into one magnificent China."[100] He distinguished it from a right-wing organization bearing the similar name of Asia Development Group (Kōa dantai), which, he claimed, had merely sought to Japanize Asia.[101]

As for the second condition, Mōri advocated the creation of a totalist state in Japan via the National General Mobilization Law. He viewed the reform of Japan's bureaucracy and national organization as having top priority. Like Okumura, Mōri called on Japan's bureaucrats to overhaul their administrative organization in order to conform to the totalist spirit of the law. He criticized the current government's liberal character and argued that implementing individual, totalist policies within such a system was pointless. As he explained, "Division and conflict within an organization inevitably disrupts and fragments a single policy, and consequently makes it unscientific and unable to display the total, scientific, planning character of totalism."[102] Mōri viewed the problem of bureaucratic reform as part of the larger problem of national organization. Effective implementation of the mobilization law depended on the active organization of the people's will. Mōri did not mention the specific form of national organization, but merely suggested that Japan's national organization should be reformed. This reform would come neither through top-down commands of an authority figure, nor disorderly chaotic change, but by recognition of the new worldview of responsible people, and the application of these principles to their actions.

97. Ibid., 28–29.

98. Ibid.

99. Originally known as the China Affairs Board (Tai-Shiin), the name was changed to Kōain in response to criticism from Japanese staff in China. See Baba Akira, *Nitchū kankei to gaisei kikō*, 350; Imura Tetsuo, ed. *Jūgonen sensō jūyō bunken shirizu.*

100. Suzuki Teiichi, "Suzuki Teiichi shi danwa sokkiroku," 114, 118.

101. Ibid., 117–119.

102. Ibid.

Mōri's National Economy

Mōri saw the China War as an opportunity for Japan to reform its economic structure and establish a so-called "national economy." In the late 1920s, Kishi had promoted the concept of the national economy as a "community of work" made possible by industrial rationalization and national cooperative effort. Combining pan-Asianism, geopolitics, and the principles of industrial rationalization, Mōri expanded the concept of national economy to justify the creation of a productivist, Asian community incorporating the economies of Japan, Manchuria, and China. As Mōri explained, "The economic space of the world is splitting apart into the space of national economies: England, America, the Soviet Union, and Germany-Italy are in the process of constructing large regional cooperative economic spheres."[103] Japan's "economic space," or "living space," had expanded as a result of the China War and the establishment of Manchukuo. He suggested that through the war "Japan's national life possesses China's living space" and therefore Japan's economy should "essentially change in order to possess China's broad economic living space."[104] In other words, the China war provided an opportunity for Japan to rationalize its economic culture—its "economic character, technology, and spirit."[105]

Mōri viewed the national economy as the antithesis of the liberal economy. He described the liberal economy as follows: the economy is based on the abstract, universal notion of *homo economicus;* capital and capital profit serve as the axes of economic life; economic activity is essentially exchange; all noneconomic elements disturb the economy's natural functioning; the economy takes precedence over politics.[106] Mōri explained that within the nineteenth century liberal world order, which he referred to as "England's national economy," abundant natural raw materials and resource-rich colonies enabled Western countries to prosper. Lacking both, Japan had to survive by purchasing cheap raw materials from abroad and processing these materials for export using its low-wage labor.[107] Under the new national economy, the main actor is not "economic man" but people who form a "concrete economy."[108] Mori appealed to the notion of an economy centered on people, not invisible natural laws. Now people become the central axes—they are no longer the object, but the subject of the economy.

103. Mōri Hideoto, "Shina ni taisuru keizai gijutsu no mondai," (Speech given at Tōa kenkyūjo on March 15, 1939). *Tōa keizai kenkyūjo* (July 1939), 12.

104. Ibid., 2.

105. Mōri, "Shina no sangyō kaihatsu," 73.

106. Kamakura Ichirō, "Nihon kokumin keizai no keisei to seiji," *Kaibō jidai* (April 1939), 26.

107. Mōri, "Shina no sangyō kaihatsu," 76.

108. Kamakura, "Nihon kokumin keizai no keisei to seiji," 26.

Drawing on Gottl's vision of an organic, rationally organized state, he suggested that laborers and entrepreneurs would acquire the status of "organizers."[109]

In concrete terms, Mōri equated the national economy with the national defense economy. He described the technical challenges of building such an economy. Japan needed to expand its total productive power to establish the East Asian New Order and mobilize a rapidly expanding defense space to prosecute the China War. Citing Friedrich List, he suggested that the new national economy of Japan must form a defense economy: "We must begin production for whatever is necessary for the security of the nation. It is meaningless to begin production of something that is merely lacking [in the world economy] but not necessary for national security."[110] Mōri listed the concrete objectives for realizing the national economy: establishment of raw material industries; establishment of machine tool and precision industries; ensuring the agricultural production of basic food stuffs; expansion and stabilization of currency and commodity exchange; improvement of the general health of the nation, including the creation of a welfare insurance system.[111] These goals reflected his concerns that the new national economy should not form a "one-sided industrial state." Within the new economy, Japanese agriculture would acquire "a new status" as would the raw material production areas of Manchuria and China.[112] The difficult challenge was to effectively tap the resources of the East Asian bloc. Mōri pointed out that although Japan now had access to a labor force of 60 million people and vast natural resources, developing these resources was expensive and reliance on locally produced raw materials such as Manchurian coal and iron would lead to higher production costs. In order to cover the higher costs and retain a leading position in the world economy, Japan would have to produce superior products based on sophisticated technology.[113]

Mōri understood the establishment of Japan's national economy as ultimately a cultural challenge. He called on Japan to foster an economic and technological culture that was suitable for Japan's expanding role in East Asia. Japan needed to transform itself from an "interest-based social organization" (*rieki shakai soshiki*) into a "community-based social organization" (*kyōdō shakai soshiki*). As he explained, through the Meiji Restoration, Japan established a modern state within the hegemonic international capitalist order and a capitalist system boldly and rapidly developed within Japan. The result was the construction of an "interest

---

109. Ibid., 30; on Gottl see Katsuta, *Nihon zentaishugi keizai no seikaku*, 16–17.
110. Kamakura, "Nihon kokumin keizai no keisei to seiji," 31.
111. Kamakura, Ichirō, "Kokumin keizai to rieki," *Kaibō jidai* (May 1939), 88.
112. Kamakura, "Nihon kokumin keizai no keisei to seiji," 32.
113. Mōri, "Shina no sangyō kaihatsu," 78.

society" (*rieki shakai*) that was "founded upon the principle of conflict" and whose actions were "prescribed by the principle of individualism." Through the China War, however, Japan's "innate totalist view" (*honraiteki zentaikan*), which he associated with "the innate, total character of the Japan of *ikkun banmin*" and "an innate, vital, existential character," was being newly awakened. This comprehensive worldview was one that "embraces all aspects of the people's livelihood and can be used as the principle of action in one's life." Such a principle would address all aspects of national life—the economy, thought, and society— and would be "synthesized three-dimensionally with the present liberal economic system": economically, through a shift from a commercial economy to a production economy; ideologically, through a shift from conceptual rationalization to historical idealism; and socially, through a shift from a conflict society to a communal society.[114]

For Mōri, the main obstacle to building a national economy was the restrictive character of Japan's existing technology, or so-called "economic technology." By the idea of "technology," or "economic technology," he meant not only technical invention, but also organization, financial mechanisms, and economic policies. Mōri called for the liberation of technology from both the domination of the economy and its "material existence" as simply an element of production and its transformation into a technology that has a "life existence" directly tied to the spiritual power of the ethnic people. He sought to remove technology from the Western liberal tradition and reunite it with Japanese culture and spirit, suggesting that "technology is not just civilization in the old sense, but can be one form of the nation's spirit, in other words, culture." Technology, which enables the transition to synthetic raw materials, was the key to the creation of an East Asian New Order because it "increases the possibility of the people's government being completely unified into an independent character and the people's government being liberated from dependence on the economy."

Mōri believed that, ultimately, technology was not a material matter, but a spiritual matter. He reasoned that since technology was the primary means to liberate Japan and Asia, it was tied to the fate of the *minzoku* and possessed a spiritual and cultural meaning. When technological innovation became the standard to measure the freedom and creativity of the Japanese people, it became a vital, state asset and could no longer be left to individual, chance discovery. Mōri believed that the fount of technological genius was ultimately human cooperative effort. He wrote: "We must seriously consider the problem of what kind of economic spirit or economic policy must be established in order to mutually

---

114. Kamakura, "Kokumin soshiki to Tōa kyōdōtai no fukabunsei," 23–27.

strengthen the Japanese economy within this stage." In terms of this spirit, he believed that it must be based on a science and technology that is not driven by profits but by the desire to advance the people's livelihood.[115]

Although Mōri's and Okumura's techno-fascist visions of a technologically liberated, functionally organized society based on the ethnic spirit may have appealed to a select group of technocrats, they were completely lost on others. On February 2, 1940, during the seventy-fifth session of the Diet, the Minseitō representative Saitō Takao launched into a devastating critique of the government's position on China. He denounced the official slogans of "holy war," New Asian Order, and "eight corners under one roof" as half-baked, empty, and highly misleading and the Asia Development Board's "Outline for a New Asian Order" as "virtually incomprehensible."[116] He questioned the wisdom of fighting a costly war with China when the government sought no real, tangible territorial or commercial gains.

Saitō's critique revealed the vast gulf that existed between the Meiji generation of leaders and supporters of the status quo and the new generation of technocrats and proponents of "reform." Saitō embraced the Meiji worldview of cooperative diplomacy, Social Darwinian "survival of the fittest," and Japan as a maritime trading partner in the world economy. Technocrats embraced a techno-fascist worldview of an organic, functionally organized Asian regional sphere in which Japan would represent Asia's headquarters as leader and producer of sophisticated technology while other countries would represent subsidiaries which would provide raw materials, labor, and other services necessary for Asia's self-sufficiency. Whereas Saitō viewed Japan's foreign policy in terms of short-term objectives based on a concrete analysis of costs versus benefits, reformists viewed Japan's policy in Asia from a long-term, one-hundred-year perspective with no expectation of immediate, tangible gains to justify the costs for both the Japanese and other Asians.

Saitō was quickly silenced and removed from the public pulpit. The Social Masses Party, under the leadership of Asō Hisashi, took the lead in demanding his expulsion from the Diet.[117] Most Diet members, sensing the political winds, either voted for his expulsion, or refrained from voting. Only a small handful of conservative Diet members voted to retain Saitō. In the coming battle over the New Order, however, influential critics in the business world and the bureaucracy could not be disposed of so easily. Instead of simply posing as the voice of

---

115. Kamakura Ichirō, "Gijutsu no kaihō to seiji," *Kaibō jidai* (September 1939), 4–5, 7.
116. See Kinmonth, "The Mouse that Roared," 334–339.
117. Ibid., 342–343.

clarity and good sense like Saitō, these leaders adopted more sophisticated and ruthless tactics to challenge and discredit the reform bureaucrats.

What was the process by which Mōri and Okumura became attracted to fascism? We can mention a number of factors that played a role. These bureaucrats developed early on a well-defined, antiliberal technocratic orientation. They became drawn to the geopolitical, pan-Asianist, socialistic ideas of right-wing ideologues, military planners, and progressive intellectuals and politicians. The desperate war situation in China and Japan's deteriorating economy encouraged them to adopt increasingly radical methods of strengthening the state and mobilizing popular support. Both were deeply impressed by Nazi Germany and sought to adapt its technological vision and political-economic policies and programs. They were also ambitious bureaucrats who saw in techno-fascism a way to enhance their own careers. It is difficult to pinpoint one particular person, group, idea, or experience which led them to embrace fascism. The point here is that their conversion to fascism did not involve one big leap, but many small, measured steps. Reform bureaucrats were pragmatic opportunists who drew eclectically on many different ideas and methods in the construction of a technocratic new order.

THE NEW ORDER AND THE
POLITICS OF REFORM, 1940–41

The New Order movement was the magnum opus of the reform bureaucrats. It represented the most direct assault on the citadels of liberal capitalism. It also represented the most ambitious bid for power by Japan's wartime technocrats. Under the leadership of Kishi Nobusuke, reform bureaucrats set out a bold and comprehensive agenda to restructure Japan's economy. Within a two-year period they drafted plans for various "new orders" in industry, finance, labor, science, technology, communications, and national land planning. The most controversial plan was the proposal to "separate capital and management" through the creation of industry-based control associations. These plans expressed a techno-fascist vision of an authoritarian managerial state in which ethnic consciousness replaced class consciousness.[1]

The zaibatsu and their supporters fought back. They accused the reform bureaucrats of being "red," which prompted the arrest of left-wing Cabinet Planning Board members and the resignation of Kishi and Hoshino. As numerous studies have argued, the zaibatsu achieved an important victory by securing control over the industrial control associations (*tōseikai*). Reform bureaucrats failed to break the power of the zaibatsu and transform Japan's economy into a unified, total war machine. If anything, the New Order hindered the mobilization effort by

---

1. The "Outline of Fundamental National Policy" was put forth on August 1, 1940. The reformist research organ, the National Policy Association, also participated in the drafting of this document. A copy of the text is contained in Kikakuin kenkyūkai, *Kokubō kokka no kōryō* (Tokyo: Shinkigensha, 1941), 18–20.

adding an additional layer of bureaucratic red-tape to the already chaotic jumble of government controls.[2]

Preparation for total war, however, was not the primary goal of the New Order. Reform bureaucrats believed that they were fighting an ideological war on two fronts: in Japan against the status quo; in China against Western liberalism and communism. They aimed at the construction of a Japan-centered Asian empire. Reform bureaucrats argued that a New Order in Japan was the prerequisite for a New Order in Asia. In order for Japan to fight a "one hundred year war" and "develop Asia" (*kōa*), it had to reform itself first.

Through the New Order, reform bureaucrats aimed to change Japan's liberal capitalist mentality and conflict-ridden culture. As reform bureaucrats made clear in Manchuria's industrialization program, they valued the expertise of Japanese business leaders but wanted them to manage their companies as public servants, not as profit-seeking capitalists. They devised new organizations, laws, and programs to recast the occupational roles and identities of managers, workers, bankers, investors, and technologists in functionalist terms. They also adopted a new rhetorical strategy. To businessmen, they appealed to the samurai spirit and urged businessmen to view their activities as a calling and a form of public service, not as a money-making enterprise. To bankers and investors, they appealed to investment patriotism and encouraged banks and stockholders to invest in the long-term future of Japan. To workers, they appealed to traditional notions of status and occupation in which all work possessed equal honor and dignity in the eyes of the state. Finally bureaucrats sought to raise the status of engineers and promote technology not as a means of production and profit, but as a national asset and expression of the Japanese spirit. This chapter examines the vision of the New Order and the battles fought over that vision.

The Shift to Wartime Controls

Once the military embarked on its program to expand munitions production, the imposition of government economic controls appeared more or less inevitable.

---

2. On the Economic New Order, see Nakamura Takafusa and Hara Akira, "Keizai shintaisei," in *Nihon seiji gakkai*, ed., *"Konoe shintaisei" no kenkyū* (Tokyo: Iwanami shoten, 1972), 71–133; Okazaki Tetsuji, "The Japanese Firm under the Wartime Planned Economy," in Masahiko Aoki and Ronald Dore, eds., *The Japanese Firm: The Sources of Competitive Strength* (Oxford: Oxford University Press, 1994), 350–378; Jerome B. Cohen, *Japan's Economy in War and Reconstruction* (Minneapolis: University of Minnesota Press, 1949); Chalmers Johnson, *MITI and the Japanese Miracle: The Growth of Industrial Policy, 1925–1975* (Stanford: Stanford University Press, 1982); Peter Duus, "The Reaction of Japanese Big Business to a State-Controlled Economy in the 1930s," *Rivista Internazionale di Scienze Economiche e Commerciali* 31, no. 9 (September 1984), 819–831.

One reason was economic: the rapid increase in military expenditure under Finance Minister Baba Eiichi brought about a balance of payments crisis that in turn called forth controls of foreign trade, finance, materials, and eventually prices.[3] The other reason was political: as Minobe Yōji put it, the military could expand munitions production and productive power either by making the munitions industries more lucrative or by controlling the production process itself. Given the reformist officers' anticapitalist stance, the former was not a politically viable option.[4] Moreover, especially in the late 1930s, there was real concern among leaders about social conflict arising from inflation and the resultant decline in the real wages of workers.[5] State planning, they hoped, would ensure that the war's costs would be more equitably distributed among the people.

Except for the controversial Electric Power Control Law, the first economic control measures were generally accepted by business as necessary measures to prepare for war. Bureaucrats presented their control policies as temporary, short-term measures to satisfy the military's growing demand for materials. The first step the government took to address the balance of payments crisis was direct control of foreign trade beginning with the Imports and Foreign Exchange Control Ordinance (*Yunyū kawase kanri rei*) in January 1937. Based on the Foreign Exchange Control Law (1933), this ordinance required the approval of the Minister of Finance on all import payments exceeding thirty thousand yen.[6]

With the inauguration of the first Konoe cabinet in June 1937, the government outlined its basic control policy in its "Three Fundamental Principles of Finance and Economics." Issued by Finance Minister Kaya Okinori and Commerce Minister Yoshino Shinji, it declared the goals of expanding productive power, conforming to international balance of payments, and adjusting supply and demand for materials. Its meaning was spelled out in a series of control measures in the following years. In September 1937 the Diet passed two important pieces of wartime legislation. The first was the Temporary Export-Import Commodities Measures Law (*Yūshutsu-nyū hinra rinji sotchi hō*), which applied across-the-board controls on all aspects of foreign trade, not only on the import and export of commodities and raw materials, but also its production, use, trade, and distribution. The second was the Temporary Funds Adjustment Law (*Rinji shikin chōsei hō*), which controlled the financing of companies including loans, issuance of stocks and bonds, increase in capital, mergers, and the establishment of new companies. The

3. Hara Akira, "Senji tōsei keizai no kaishi," *Nihon no rekishi*, vol. 20, *Kindai* (Tokyo: Iwanami kōza, 1981), 218–268.

4. Minobe Yōji, "Keizai shintaisei no igi to kōzō," in Minobe Yōji, ed., *Senji keizai taisei kōwa* (Tokyo: Tachibana shoten, 1942).

5. Okazaki, "The Japanese Firm under the Wartime Planned Economy," 362.

6. This sum was eventually reduced to 100 yen.

law classified these companies in terms of their importance to munitions production, with military-related industries receiving abundant funds.

War mobilization reached a new stage when the state began to control access to and distribution of materials in the fall of 1937. The Cabinet Planning Board drafted the first of a series of materials mobilization plans (*Busshi dōin keikaku*) to allocate basic materials among the army, navy, and civilian sector. The central variable in materials planning was import capacity because of the country's heavy dependence on foreign trade. Based on an assessment of total foreign currency reserves and the domestic supply of commodities available, planners determined the type and quantity of commodities to be imported. Implementation of the mobilization plans was carried out through the Commerce Ministry's Temporary Materials Coordination Bureau (Rinji busshi chōseikyoku), which sought to control distribution, consumption, and production of these materials through negotiation with other government agencies and by appealing to industrial "self-control." At the same time, the Cabinet Planning Board began drafting the Production Capacity Expansion Plan (*Seisan ryoku kakujū keikaku*) as the domestic component of the Manchurian Five Year Plan. It postponed adopting the plan until 1939, however, and changed them to annual plans for expansion of productive capacity based on a four-year plan to increase production in key heavy industries.[7] With the gradual shift from a price-based to materials-based economy, these plans formed the cornerstone of the war economy.

Despite the totalitarian reach of the National General Mobilization Law (*Kokka sōdōin hō*), its enactment was a relatively uncomplicated affair. In contrast to the Diet's extended negotiations over the Electric Power Control Law, it drafted, deliberated, and passed the national mobilization law within half a year. Whereas critics denounced the Electric Power Control Law as a political vehicle to promote fascism, the mobilization law was seen as a provisional and necessary war measure. Prime Minister Konoe assured the Diet that the law would be used only during a state of war, which did not include the China Incident. The law was composed of fifty articles covering all aspects of economic activity including industry, finance, labor, education, and the media. It granted the government sweeping powers to control most aspects of the economy through imperial ordinances for the purposes of war.

The complacent attitude toward the mobilization law was soon reversed. In November 1938, bureaucrats invoked Article 11, which permitted the state to

---

7. For a detailed examination of the planning process, see Okazaki Tetsuji, "The Wartime Institutional Reforms and Transformation of the Economic System," in Banno Junji ed., *The State or the Market? vol. 1, The Political Economy of Japanese Society* (Oxford: Oxford University Press, 1997), 277–345.

intervene in the financial management of firms, including the disposal of profits, issuance of bonds, and management of companies. Bureaucrats wanted to control the dividend rates and surplus profits of companies, which were making handsome profits from the war.[8] Opposition to Article 11 was led by Ikeda Seihin, a Mitsui executive who held the ministerial portfolios of both finance and commerce. Ikeda and the business community warned bureaucrats that state control of profits would severely impact business performance.

The Company Profits, Dividends, and Funds Accommodation Ordinance in April 1939 represented a victory for the business community. The new ordinance only applied to companies capitalized at 200,000 yen or more. It merely limited regular dividends to 10 percent and required firms to retain excess profits. The ordinance did not require companies to invest their profits into the war effort or development projects in China. The Ministry of Finance chose a less confrontational strategy to raise capital through the Industrial Bank of Japan, which would use its expanded powers and influence over firms to raise capital for projects that the state considered essential for the war effort.[9]

The successive appearance of new planning organs and control measures, coupled with Konoe's proclamation of a New Order in Asia, raised concerns that the government's control policies indicated something more permanent. The Ministry of Commerce created a variety of external bureaus, including the Temporary Industrial Rationality Bureau, Temporary Materials Coordination Bureau, Fuel Bureau, and Trade Bureau. The new Welfare Ministry created a Labor Affairs Bureau and the Finance Ministry established the Foreign Exchange Bureau and Funds Bureau.[10] In December 1938, leaders created the Asia Development Board to coordinate China policy.

Reform bureaucrats, who staffed the new planning apparatus, savored their newly acquired authority. Sakomizu Hisatsune, who drafted the wartime financial control policies as Finance Section chief of the Finance Ministry's Public Finance Bureau (Rizaikyoku), described control as "humans pretending to be gods."[11] Through the Temporary Funds Adjustment Law, he and his colleagues decided the fate of a company. According to Sakomizu, his section served as the "window of the Finance Ministry on all things concerning the problems of control and the main contact to the Cabinet Planning Board."[12] Minobe Yōji, after

---

8. Miriam Farley, "The National Mobilization Controversy in Japan," *Far Eastern Survey* 8, no. 3 (February 1, 1939), 28.

9. Sakomizu Hisatsune, *Kaisha rieki haitō rei gaisetsu* (Tokyo: Ōkurashō, 1939).

10. Okumura, *Nihon kakushin no seiji*, 115.

11. Sakomizu Hisatsune, Interview, in Nakamura Takafusa, Itō Takashi, and Hara Akira eds., *Gendaishi o tsukuru hitobito,* vol. 3 (Tokyo: Mainichi shinbunsha, 1971), 70.

12. Ibid.

returning from Manchuria, held a variety of posts in the Commerce Ministry's Temporary Industrial Rationality Board, Industrial Affairs Bureau, Temporary Materials Coordination Bureau, and Price Bureau. He became identified with the state's control of the textile industry through the so-called "link system," in which imports of raw cotton were linked to exports of cotton products. Minobe's name became synonymous with "textile control."

When Konoe announced the New Order movement in July 1940, the public reacted with bewilderment and apprehension. Critics asked why the government was creating more uncertainty and unease in an already chaotic situation. Few understood the purpose of Konoe's New Order. In one roundtable, an Osaka businessman was quoted as suggesting that the New Order would enable Japan to quickly conclude the fighting in China and return to a liberal capitalist economy.[13] Another suggested that the New Order was an attempt to address the shortcomings of the controlled economy. Others interpreted it as primarily a military response to external developments and a plan to build a national defense state based on the lessons of the European war. One of the main reasons for the confusion about the New Order was that it marked a radical departure from the public's assumption that economic controls were part of the military's plan to build a national defense state for total war.[14] The New Order movement confirmed suspicions among business that reform bureaucrats were building a fascist state and an Asian empire.

From Total War to Totalist War

The military's conception of the national defense state was based on the vision of total war and the state's central role in mobilizing the nation's resources for such a war. From the perspective of total war theorists such as Ishiwara Kanji, the China War was a deviation from the original plan to develop the nation's resources for a "final war" against either the Soviet Union or United States. It served no constructive purpose other than to divert precious resources, as well as foster anti-Japanese sentiment in China and the outside world. As Japan became increasingly entangled in the battles in China, leaders turned to new justifications for war.

Konoe's proclamation of a "New Order in East Asia" following the fall of Wuhan and Guangzhou in the autumn of 1938 represented the first tentative attempts to

---

13. Shiina Etsusaburō, Kōno Mitsu et al., "*Shinataisei to kōgyō no saihensei*" *o kataru zadankai*, *Kōgyō kumiai* (October 1940), 31.
14. Hashikawa, "Kokubō kokka no rinen," 246.

promote a new interpretation of war and a new worldview. Leaders portrayed the China War as an ideological war, not as a military war for resources. In their writings and speeches from this time, reform bureaucrats and their civilian supporters interpreted the China War as a historical turning point and transition to a "totalist" world order. Four months after the Marco Polo Bridge skirmish in July 1937, Okumura proclaimed that the China Incident was more than just a clash of military forces between China and Japan, but rather an expression of the greater transformation taking place within the world and one sign of the coming second world war between the "have" and "have-not countries," and between liberalism and totalism.[15] China served as the battlefield on which Japan was fighting the "have-countries" (*moteru kuni*) England and the Soviet Union.[16] He referred to the battle as merely the first stage of a "one hundred year war" to build an "Asian cooperative body" or an "Asia for Asians."[17]

Reform bureaucrats argued that in order for Japan to win a totalist war, it had to build a totalist advanced national defense state. In posing the question "What is a national defense state?" Okumura explained that it was neither a state that emphasized national defense, nor one in which the military controlled every aspect of the state. "The national defense state must be understood not on the basis of the common notion of the national defense state, but on the basis of a new state view that is different from the liberal conception of the state up until now."[18] He noted that the meaning of "reform" did not exist apart from the creation of an advanced national defense state. The implementation of reform, creation of the New Order, and completion of the national defense state system were one and the same.[19]

According to Okumura the national defense state developed in three stages.[20] In the first stage, the state responded to the need for the rapid expansion of military goods through "general, mechanical, liberalistic mobilization." This "mechanical, low-grade mobilization" took the form of expansion of munitions production in factories through control over munitions-related industries, funds, raw materials, labor, and consumption, and eventually over trade. In the second stage, the state progressed toward an advanced national defense system. The state continued to pursue similar controls as in the first stage, but now more efficiently,

---

15. Okumura Kiwao, "Tōa kensetsu hyakunen sensō," in Okumura, *Henkakuki Nihon seiji keizai*, 144.

16. Okumura, "Tenkanki ni tatsu sekai to Nihon," in ibid., 133–134.

17. Okumura, "Tōa kensetsu hyakunen sensō," 145; Okumura Kiwao, "Tōsei kokka no kansei to eirishugi," in Okumura, *Henkakuki Nihon seiji keizai*, 151.

18. Okumura Kiwao, "Kokubō kokka toshite no Nihon," in Okumura, *Nihon seiji keizai*, 332.

19. Okumura Kiwao, "Kakushin to wa nanizoya," *Jitsugyō no Nihon* (January 1941), 44, 46.

20. Okumura, *Nihon kakushin no seiji*, 107–114.

resulting in the development of the profit-based, liberal economy to an advanced degree. During this stage, the contradictions of the liberal capitalist economy became pronounced. The demand for the rapid expansion of munitions brought about the trend toward large-scale production and the destruction of small- and medium-sized companies and nonmilitary related industries. The results were greater controls over labor, inflation, and the impoverishment of the masses. Eventually the advanced national defense system became deadlocked and unable to respond to the demands of the state and people. The third stage represented the "reform" and "refortification" of the advanced national defense state through the replacement of individualism with totalism, and the liberal capitalist controlled economy with a totalist planned economy. A new form of state monopoly capitalism based on totalist thought would replace private monopoly capitalism.

Reform bureaucrats believed that Japan had reached the end of the second stage and was beginning to exhibit the worst characteristics of both the liberal and controlled economy. Material and price controls had brought about a decline in both production and the circulation of goods, giving rise to hoarding and a flourishing black market. Bureaucrats faced mounting criticism from the private sector. Minobe Yōji admitted that their control techniques were clumsy. But the more fundamental problems were of a political, not technical nature. Facing resistance from business, the government had left the previous liberal economic structure intact and had tried to steer the economy from above through legal and administrative measures. As the government gradually extended its authority down to the smallest detail, it deprived the national economy of its vitality, which in turn brought about conflicts between controller and controlled and estrangement between bureaucrats and the people.[21] As Mōri Hideoto saw it, the economy had not progressed beyond an "administered" or "guided" controlled economy. The state still operated within the liberal economic system and applied "administrative techniques," "legalistic controls," and "stop-gap measures" to individual economic phenomenon. The controlled economy lacked both a true planning character and a clear political will with the result that administration conflicted with national livelihood and led to its stagnation and the loss of its "self-regulative and autonomous character." What was needed, Mōri argued, was the establishment of a "basic political will" and a "new organization of national life."[22] Without a strong government and national party, a unified controlled economy could not be realized.

---

21. Minobe Yōji, "Keizai shintaisei no igi to kōzō," in Minobe Yōji, *Senji keizai taisei kōwa* (Tokyo: Tachibana shoten, 1942), 21.

22. Kamakura Ichirō, "Tōsei keizai no hinkon no genin: Shizenryoku ka soshiki ryoku ka," *Kaibō jidai* (December 1939), 18–21.

The Limits of Political Reform

Prime Minister Konoe was a useful figure to both the military and to civilian reformists. The army saw Konoe as the only suitable leader who could push through its domestic agenda.[23] Reformists relied on Konoe's authority and prestige to promote their fascist agenda. Konoe appointed reform bureaucrats such as Kishi and Hoshino to ministerial positions after they returned from Manchuria. He also lent his prestige to the leading private technocratic think tank, the Shōwa Research Association. A descendent of the ancient Fujiwara line and a confidante of the emperor and the last surviving Meiji statesman, Saionji Kinmochi, Konoe had impeccable credentials. He also enjoyed the broad support of many groups on the right and the left, including military officers, business leaders, party politicians, scholars, and journalists.[24] Kishi described Konoe as a "sympathetic politician who could understand the feelings of the younger generation."[25]

In the postwar period, Konoe's supporters defended his actions by arguing that he had sought to restrain the military. The political scientist Yabe Teiji, who was a member of the Shōwa Research Association and a close associate of Konoe, argued that Konoe was concerned that an uncontrolled military would prevent the cabinet from carrying out its duties and bringing an end to the war in China.[26] If the military was permitted to continue unchecked, he argued, it would have established a Nazi-style party dictatorship that would control thought and politics, strip the Diet and parties of any influence, and manipulate the emperor. According to Yabe, Konoe believed that strong government leadership based on a national party with popular support was the only means to restrain the military. He argued that Konoe's main goal was to recover the prestige and authority of the cabinet and Diet, overcome the intense rivalry among factions, and formulate a unified response toward domestic and foreign policy.

Yabe's account was highly misleading because the leading proponents of Nazism and Fascism during the early 1940s were civilians, not the military. When Konoe began to explore the possibility of forming a national party in the spring of 1938, he turned to Minseitō and Seiyūkai leaders, the Japanist right wing, and civilian reformists. Reformist plans were drafted by Social Masses Party leaders Asō Hisashi, Kamei Kan'ichirō, and Akamatsu Katsumaro in consultation with other

23. Itō Takashi, "*Konoe Shintaisei*" *no kenkyū* (Tokyo: Iwanami shoten, 1972), 5–7. See Yabe Teij, "Konoe Fumimaro to shintaisei," *Kindai Nihon o tsukatta hitobito* (Tokyo: Mainichi shinbunsha, 1965), 92.

24. Yoshida Tokujirō, "Konoe no 'bureen torasuto,'" *Chūō kōron* (July 1937), 274–275.

25. Kishi Nobusuke, Yatsugi Kazuo, and Itō Takashi, "Shōkō daijin kara haisen e," *Chūō kōron* (October 1979), 290.

26. Yabe, "Konoe Fumimaro to shintaisei," 92.

reform bureaucrats and army planners such as Mōri, Okumura, Katakura, and Mutō. The group proposed a radical plan to dissolve the existing parties and create a "Great Japan Party Unit" (*Dai Nihontō bu*). Kamei and Akamatsu believed that in order for a party to mobilize the political energies of the nation, it must inspire the people and provide them with a "dream" or "concrete worldview."[27] They advocated a fascist party modeled on the Concordia Association and Nazi Party. The party would promote the pan-Asianist idea of an East Asian body and encompass China's national organization, the Concordia Association, and a new national political organization in Japan.[28] The vanguard unit would oversee the party's platform, policies, and administration, and serve as an intermediary organization between the emperor and Japan's national organization. Yabe and his colleagues at the Shōwa Research Association also proposed radical plans for a greatly emaciated Diet, a mass party composed of vocational groups, and a vanguard unit based on the Nazi concept of "one nation, one party."[29] Konoe did not embrace these radical plans, however, and opted for more moderate solutions.

Unlike Nazi Germany and Fascist Italy, which directly mobilized popular support through a mass party, Japan was constrained by its conservative imperial system. That system was designed to protect imperial prerogative and enhance the power of Meiji oligarchs at the time of the founding of Japan's modern state. Government regulations restricted the sphere of political activity to the Diet, required official approval for political gatherings, and prohibited civil servants, members of the military, women, religious leaders, students, and teachers from engaging in politics. As the historian Gordon Berger pointed out, "political mobilization" was associated with "private," "partisan" interests, whereas "public mobilization" was associated with "transcendental," "national interests."[30] The officially sanctioned means to promote political agendas, then, was through public mobilization in the form of moral suasion and ostensibly for the purpose of defending the imperial system and the institutions supporting it.[31]

Working within the status quo, the political parties, right-wing groups, and Home and Education ministries promoted various proposals for public mobilization. Konoe tried to incorporate these proposals into the Political New Order. Launched in August 1940 following the dissolution of the political parties, the

27. Kamei Kanichirō, Akamatsu Katsumaro, et al., "Shintaisei no senro o kataru," *Nihon hyōron* (December 1940), 102.

28. Itō, "*Konoe Shintaisei*," 79.

29. Fletcher, *Search for a New Order*, 139–141.

30. Gordon Berger, *Parties Out of Power in Japan, 1931–1941* (Princeton: Princeton University Press, 1977), 178–184.

31. Sheldon Garon, *Molding Japanese Minds: The State in Everyday Life* (Princeton: Princeton University Press, 1997).

Political New Order movement advanced four objectives: the harmonization of the supreme command with national affairs, promotion of a unified and efficient government, establishment of a so-called "Diet assistance system," and formation of a national organization. On October 12, 1940, Konoe settled on the Imperial Rule Assistance Association. The new political organization was an unwieldy, heavily bureaucratic, public organization that lacked centralized leadership and a distinct mandate. The Home Ministry quickly appropriated its regional branches into the ministry's own administrative network.

Reformists were sharply critical of the association's vague platform and organization and its attempts to cater to all interests. They pointed out that the four-part agenda advanced by Konoe represented four bodies, not four aspects of one body, and failed to represent "one organic body in which blood circulates throughout the four organs."[32] The Political New Order was a far cry from the vision of a dynamic, advanced national defense state that could generate new energy and inspire the people. It sought only to foster passive compliance through the traditional appeals to imperial duty, self-sacrifice, and hard work. Since the prospects for political reform were bleak, reformists focused their energies on creating a new economic system. Under their plans for an Economic New Order, reform bureaucrats hoped to mobilize the people through new organizations and a new vision.

Planning the New Order

In trying to convey the spirit of the movement for a new economic system, a colleague of the reform bureaucrats described it as a bold attempt by young, earnest bureaucrats and military men to achieve "something new" by trying to directly address the root causes of national weakness and social injustice brought about by the liberal system.[33] The New Order movement was unprecedented in the comprehensiveness and breadth of its vision. Similar to Manchurian industrialization, the New Order represented a planning challenge to build a new economic system according to a series of integrated plans, programs, and institutions and to design rational new orders in every aspect of society.

At the heart of the New Order movement was a new social vision. Reform bureaucrats advocated a change from a "proprietary society" (*shoyū shakai*), driven by capital ownership and the pursuit of private profits, to a "functionalist society" (*shokunō shakai*), in which people served the state through their respective

---

32. Kikakuin kenkyūkai, *Kokubō kokka no kōryō*, 38.
33. Nihon hyōron shinsha, ed. *Yōyōtaru*, 171–173.

occupations. In contrast to the liberal capitalist vision of a class-based society, the new vocation-based society promoted the separation of capital and management and the elevation of the status of managers, technologists, and labor within society. As Mōri explained:

> We are demanding the organization of the vital duties of the whole of national life on a functionalist basis instead of abstracting politics, culture, or economy from national life...we affirm the diverse and vital functions of the people. These vital functions, however, must be organized from the standpoint of the part vis-à-vis the whole of national life. We must clarify the meaning of the vital functions of national life as a whole such as those possessed by capitalists, entrepreneurs, managers, and workers and which are strictly limited within the capitalist economy. We emphasize that, in this sense, capitalists, entrepreneurs, and workers must be organized on a functionalist basis within their present place of employment. We view it as an organization that develops and organizes within the individualist order of the Japanese people in order to develop the Japanese people today and revive its historical importance.[34]

Kishi bemoaned the tendency in Japan toward conflict, which existed in every aspect of industry—between industrial associations and trade associations, villages and cities, large industries and small- and medium-sized industries, producers of raw materials and manufactures, and producers and distributors. In one roundtable on the New Order he stated: "We need to eliminate the friction caused by the conflict-ridden, profit-loss relations we have seen in industrial economic matters up until now and create a system which goes beyond conflict to integration, one that integrates all the forces into one." What was needed was a sense of "joint public profit" (*kōkyō no rieki*) or "public profit" (*kōeki*).[35] Kishi believed that when private industry developed a sense of public interest, the nature of its relationships would change from one of conflict to cooperation. In such a society, there were no winners or losers, rather all possessed equal status and pooled their efforts together to improve the general welfare.

At the same time, reform bureaucrats acknowledged the central role of bureaucrats within the new functionalist society. Defending themselves against accusations of bureaucratic high-handedness, they called for "unity between bureaucrats and civilians" based on the mutual recognition of their distinct roles.

---

34. Kamakura, "Tōsei keizai no hinkon no genin," 20.
35. Kishi Nobusuke, "Keizai shintaisei to wa nani ka," in Kishi Nobusuke, *Nihon senji keizai no susumu michi* (Tokyo: Kenshinsha, 1942), 124–125, 126.

With regard to the control associations, Kishi suggested that "officials should naturally demonstrate their function as bureaucrats, and civilians should completely display the functions they set out to achieve as civilians." In other words, he explained, input should be provided by civilians and total planning directed by bureaucrats; officials should be charged with drafting plans and civilians with executing production. Only then will plans and results be correlated and problems such as low efficiency be overcome.[36] Sakomizu compared bureaucrats to leaders on a bus who get passengers (civilians) to make room for incoming passengers and reprimand those who are out of line.[37]

The Economic New Order for Industry

The movement for an "Economic New Order" (*Keizai shintaisei*) was somewhat of a misnomer because it dealt primarily with the reform of industry, especially its management. Its goal to "separate management from capital" formed the central plank in the movement to create a functionalist society. The Economic New Order was directed by the core group of reform bureaucrats who had designed Manchuria's planned economy. The logical coherence and unity of vision in the new order plans for industry, finance, science-technology, and labor were facilitated by the close personal networks developed among reformists through their shared experience of planning and implementation in both Manchuria and Japan.

Kishi, who returned to Japan as vice minister of commerce in 1939, spearheaded the Economic New Order. He served under the new commerce minister, Godō Takuo, a navy technocrat who had worked in Manchuria. Kishi turned down the offer of president of the Cabinet Planning Board and recommended his Manchuria colleague Hoshino instead.[38] Kishi called back Shiina Etsusaburō from Manchuria and made him director of the Commerce Ministry's General Affairs Bureau (Sōmukyokuchō), one of the most powerful positions within the control apparatus. At the Cabinet Planning Board, young planners from the various ministries drafted the new orders. The key strategies and concepts were formulated at the so-called Cabinet Planning Board Deliberation Room by reform bureaucrats such as Minobe Yōji, Mōri Hideoto, Sakomizu Hisatsune, Okumura Kiwao, Kashiwara Heitarō, Yamazoe Risaku, and Murata Gorō. Under Colonel Akinaga, who was appointed research official in the first department, Minobe,

---

36. Ibid., 130–131.
37. Kashiwara Heitarō, Minobe Yōji, Mōri Hideoto, and Sakomizu Hisatsune, "Zadankai: Kakushin kanryō—shintaisei o kataru," *Jitsugyō no Nihon* (January 1941), 56–57.
38. Kishi, Yatsugi, and Itō, "Shōkō daijin kara haisen e," 290.

Mōri, and Sakomizu formed a planning triumvirate. They served as Akinaga's key advisors, or "three ravens" (*sanba garasu*), with each assuming a particular function: Mōri came up with the basic concepts, Sakomizu arranged them into coherent policies, and Minobe oversaw their implementation.[39]

The New Economic Order aimed to reform the management strategy of firms. It sought to change the goal of business leaders from maximizing profits to increasing production of goods vital to the state. Reformists believed that the main problem was to separate the close ties between capital and management. If managers were relieved from the pressure to produce profits and pay dividends, they would direct their energies and capital to increasing production and serving state goals. Kishi set out three tasks for bureaucrats.[40] The first task was to change the form of business enterprise through plans, organizations, and laws. The second task was to foster a new public spirit among business leaders. The third and most difficult task was to use this spirit to obtain the desired results.

One of the first measures undertaken by bureaucrats was to strengthen the state's control over profits. In October 1940, they passed the Ordinance for Control of Corporate Finance and Accounting. In contrast to the previous ordinance issued under Article 11 of the mobilization law, the new ordinance expressed a clear ideological intent to redefine the nature of the firm. According to Sakomizu, originally bureaucrats had proposed to change the Commercial Code, which defined the corporation as a profit-making entity. In the end they decided to take a less confrontational route and redefine the role of profit. They argued that "profit itself is not the goal, rather by bearing the responsibilities assigned by the national economy to achieve state goals, profit as the true principle of firm management emerges as a result."[41] Reflecting the new meaning of profit, the ordinance effectively lowered the regular dividend rate to 8 percent and added the proviso that rates from the previous fiscal year could not be exceeded. Bureaucrats hoped that, by further capping dividend rates, they would reduce speculation in stocks. Sakomizu explained: "Because the dividend rate would gradually change to a fixed interest rate, the stock's yield and price would stabilize. So-called stocks would assume more the character of a fixed-interest bearing security."[42] Similar to the previous ordinance, it permitted the government to intervene in the accounting practices of firms and restrict compensation. It also limited the bonuses of company directors to a certain percentage

---

39. Nihon hōron shinsha, ed., *Yōyōtaru*, 180, 318, 343.
40. Kishi Nobusuke, Shiina Etsusaburō et al., *Shin keizai taisei no kakuritsu o kataru zadankai sokkiroku* (Tokyo: Sekai keizai shinposha, 1940), 7–11.
41. Sakomizu Hisatsune, "Kaisha keiri tōsei rei o kataru," *Jitsugyō no Nihon* (April 1940), 14.
42. Ibid., 17.

of profits, thereby weakening the tie between executive compensation and firm profits.[43]

More radical measures were in the pipeline, however. In their first drafts for the Economic New Order presented in September 1940, reform bureaucrats proposed a reorganization of companies into a three-tiered system composed of a government supervisory organization at the top, sector-based control organizations, and member firms.[44] The heart of the new system was the industry and materials-based control organizations. Originally called "national production cooperative bodies," their name was changed to "control associations." They were defined as public entities that represented a particular industry or material. Large firms were obligated to join one or more control associations depending on their area of business. Smaller firms were forced to merge with the large firms. In addition, all business councils, trade associations, and syndicates were required to join the control associations. Control associations were placed under the direct control of a higher, unifying organ referred to as the "central economic headquarters." This supervisory body participated in drafting various state plans by providing basic information concerning industries and materials. It advised the government and represented the control associations in their negotiations with the various ministries. In the first drafts of September and October 1940, the government would appoint the leader of the supervisory organization.

Bureaucrats sought to separate capital from management by placing firms under the control of a higher authority embodied by the control associations and the government's supervisory body. The head of each control association assumed the official position of *Führer* and exercised near-dictatorial powers over member firms. The term *Führer* was directly taken from the Nazi concept of one who assumes dictatorial powers over the firm and full responsibility for its employees and the general public welfare. The leader exercised authority over all matters concerning that particular industry. He determined the type, price, quantity, as well as method of production and distribution of goods. In consultation with the state, he supervised the member firms' financial accounting, profits, compensation policy, science and technology research, as well as the hiring and dismissal of directors of member firms and their suborganizations. The control associations played an important intermediary function in coordinating production commands from the government with implementation of the plans among firms. According to Okazaki Tetsuji, under the "rigid imperative planning model," information was not conveyed via price, but via the downward flow of

43. Ibid., 21–22; Okazaki, "The Wartime Institutional Reforms and Transformation of the Economic System," 296.

44. Okazaki, "The Wartime Planned Economy," 367.

commands from the top. The control associations performed an important "iterative function" by participating in both government planning and in allocating and adjusting orders to members firms.[45] Under this system, stockholders had little say in the management of their firms.

The concept of the control associations was informed by various managerial concepts and models advanced in the United States, Nazi Germany, the Soviet Union, and Manchukuo. The idea of "separating management and capital" was already well known in Japan through the famous study of Berle and Means and Okumura's draft of the electric power law, and was popularized in Ryū Shintarō's bestseller, *The Reorganization of the Japanese Economy.*[46] The idea of the "cooperative body" drew on German concepts of industrial rationalization, Gottl's "ethnic cooperative body," and Nazi concepts of *Volksgemeinschaft, Führerprinzip, and Gemeinnutz vor Eigennutz.*[47] Finally, Manchuria's special companies provided a firsthand model of industry-based control associations.

In order to change the spirit of management, bureaucrats felt it necessary to change the occupational status and identity of business leaders. Control association leaders were given official status as economic bureaucrats. In a country such as Japan which exalts officialdom, conveying bureaucratic status represented one of the highest forms of social recognition. As the Shōwa Research Association explained in their draft of the New Order, managers would serve two roles: to produce a fixed amount of profits to shareholders and, more important, to fulfill their public duties to increase production. As public officials, they would be motivated to display their creativity and ability to the greatest degree.[48] Kishi called for a new business ethic that reflected the public character of the firm and its management. He explained that the nature of business enterprise would change from the pursuit of profitable investments to providing a commission-based service. Kishi exhorted leaders to view their work as a vocation or calling. He appealed to the dignity of business leaders and their moral superiority, samurai spirit, and patriotic fervor.[49]

The draft of the New Economic Order was denounced by business leaders and the new commerce minister, Kobayashi Ichizō, a powerful Osaka industrialist. Business leaders accused the reform bureaucrats of promoting communism. Opponents of the Economic New Order delivered two blows to the reformists. Both Kishi and Hoshino were forced to step down from their posts at the Commerce

---

45. Ibid.

46. Fletcher, *Search for a New Order,* 127–129.

47. In a confidential Cabinet Planning draft of the Economic New Order, Gottl's name is mentioned. Minobe Yōji Documents: G 2:2, 2:5, 2:6.

48. Shōwa kenkyūkai, "Nihon keizai saihensei shian," 18–19.

49. Kishi Nobusuke, "Atarashiki shōnindō," *Yūben* (November 1940), 28.

Ministry and Cabinet Planning Board. From January until April 1941 seventeen members of the Cabinet Planning Board were arrested on the grounds of violating the Peace Preservation Law. Among those arrested were prominent postwar socialist leaders, Wada Hiroo, Inaba Hidezō, and Katsumata Seiichi.

By the time the control associations were established in August 30, 1941, business had succeeded in taking the teeth out of the original plan and seriously compromising its vision. The most important concession to business was to permit them to elect the various leaders of the control associations. The führers were inevitably the presidents and directors of the most powerful firms, syndicates, and trade associations. They proceeded to incorporate their existing, profit-oriented cartels into the new structure and, through them, strengthen their own firm's control over smaller firms through their new powers to "rationalize" industry. Big business conducted business as usual, though now their powers were more concentrated and backed by the state.

The New Order for Finance

Reform bureaucrats formulated measures to not only separate management from capital but also to change the role of capital and finance. The goal of the New Order for Finance was to encourage owners and managers of capital to adopt a production orientation by supplying funds to priority industries and taking a longer-term investment perspective. More generally, they sought to change the nature of finance from an independent, profit-making business to a public service enterprise.[50]

In the fall of 1940, Finance Ministry bureaucrats drafted the "Outline of Basic Fiscal and Financial Policy." The outline proposed top-down planning of the state's financial resources and the mobilization and distribution of public and private sector funds. The New Order for Finance was based on a State Funds Mobilization Plan (*Kokka shikin dōin keikaku*) to determine the overall distribution of total funds for civilian consumption, public finance, and industry.[51] Under this plan, funds for industry were allocated and regulated and a separate funds distribution plan for each industry was created. Sakomizu, who helped draft the plan, explained that the main thinking behind the plan was that state funds as a whole should be managed from the perspective of the production cooperative

---

50. Sakomizu Hisatsune, *Zaisei kinyū kihon hōsaku kaisetsu* (Tokyo: Senji seikatsu sōdansho, 1941), 31.

51. Since 1939, annual Funds Control Plans (Shikin tōsei keikaku) were drafted by the Cabinet Planning Board as a component of materials mobilization planning to plan for the distribution of funds on the consumption side.

body as a whole and not from the dual perspectives of public and private finance. He noted that under the new system, the Finance Ministry would wield unprecedented power, becoming "the accounting section head of the Japanese state, which is the so-called production cooperative body."[52]

The main institutional innovation was the move toward a system of "cooperative financing" among private banks. The concept of cooperative financing or cooperative loan syndicate was advanced primarily in response to the diminishing role of the stock market in financing companies during the wartime period. Under the peacetime liberal economy, the stock market had served as the main vehicle for firms to raise funds. During the early 1930s firms obtained approximately 30 to 40 percent of their funding through the issuance of stocks. By the early 1940s, this percentage had dropped to below 20 percent.[53] Under the liberal system, profitable firms with rising stock prices were able to raise capital by issuing stock. Moreover, profitable firms that were primarily financed by equity and had a low debt-to-equity ratio were able to obtain loans from banks more easily and cheaply because of the higher expected probability of loan repayment. With the shift to a controlled economy and the restrictions placed on dividend rates, prices, and management, firms became less profitable and less attractive investment vehicles for stock investors and banks.

In order to encourage investment in firms in military-related industries, which often had higher risks and were less profitable, planners sought to change the lending relationship between firms and banks. In October 1940, the government passed the Banks and other Financial Institutions Funds Utilization Ordinance (*Ginkō to shikin unyō rei*), which enabled it to order financial institutions to lend funds to firms in priority industries. In place of lending by individual banks based on free choice and the criteria of loan repayment, the government now assigned banks to specific firms in priority industries as "appointed banks," or "main banks."[54] Banks were now obligated to make loans to assigned firms and monitor their activities. At the time, banks were ill equipped to carefully monitor firms under their care and to scrutinize loans.[55] According to Okazaki, major banks lacked sufficient capabilities both in independent credit analysis to effectively assess the increased risks and greater volume of lending and in organizing syndicated loans. As a result, the government stepped in to organize loan consortia under the guidance of the Industrial Bank of Japan, which had

52. Sakomizu Hisatsune, "Kokka sōryoku sen to zaisei," *Tōyō keizai shinpō* (April 1943), 78.

53. Okazaki, "The Japanese Firm under the Wartime Planned Economy," 351–352.

54. Okazaki Tetsuji, "Wartime Financial Reforms and the Transformation of the Japanese Financial System," in Erich Pauer, ed., *Japan's War Economy* (London: Routledge, 1999), 156.

55. Itō Osamu, "The Transformation of the Japanese Economy," in ibid., 174–175.

extensive experience in financing the government's high-risk munitions and overseas investments as well as the new zaibatsu. In October 1941 the Industrial Bank of Japan and ten major city banks created the Emergency Cooperative Loan Syndicate (Jikyoku kyōdō yūshi dan) to cooperatively lend to firms. Within the consortia, the Industrial Bank of Japan and the firm's main bank were responsible for monitoring the firm's loan. In 1942, this loan syndicate was replaced by the National Finance Control Association and the supervisory role of the Industrial Bank of Japan was overtaken by the Bank of Japan.[56] Through the new system of cooperative financing, the government was able to ensure a steady supply of funds to firms, which were consequently released from pressure to maximize short-term profits to obtain funds, and financial institutions were more willing to lend to firms in high-risk priority industries. Both capital and management could shift their orientation from one focused on commercial profits, to one focused on production for the state.

Reformists also sought to change the goals of stock investment and eradicate the speculative nature of the stock market. They appealed to stockholders to adopt "investment patriotism" and change their attitude toward investing from one focused on short-term profits to one that "takes the long and broad view of Japan's national power."[57] Sakomizu urged investors to base their investment decisions on considerations of state objectives: "I would like capitalists to think about the profits of the country as a whole rather than individual profits, to take the long view and be prepared to fully contribute to the development of the expansion of production of the country, and to invest accordingly."[58]

Bureaucrats adopted measures to control stock prices and thereby reduce investment risk in strategic industries. In April 1941, they established Japan Cooperative Securities (Nihon kyōdō shōken) with funds provided by banks, insurance companies, other financial agencies, and large corporations. The purpose of the company was to increase investor confidence in the stock market by purchasing shares in the case of sharp price declines and to assume the portion of new shares not underwritten in the case of stock issuance. The activities of Japan Cooperative Securities were supported through the Stock Price Control Ordinance (*Kabushiki kakaku tōsei rei*), which empowered the finance minister to set minimum stock prices in special cases in which there was concern that the company was unable to stabilize prices. The ordinance represented a type

---

56. Okazaki, "Wartime Financial Reforms and the Transformation of the Japanese Financial System," 156–165.

57. Sakomizu Hisatsune, "Senjika no kabushiki mondō," *Jitsugyō no Nihon* (November 1941), 28.

58. Ibid., 33.

of insurance policy to prevent the government from having to suspend or close the market in the case of steep declines. In addition, the government provided further incentives for holding stocks through the Temporary Measures for Valuation of Company Stockholdings Ordinance (*Kaisha shoyu kabushiki hyōka no rinji sochi rei*). Under this ordinance, firms were relieved from having to report unrealized stock losses in order to encourage long-term holding of stocks. The new ordinance essentially represented a revision of the Commercial Law, which stipulated that a firm's securities be valued based on the current market price. Now stocks held by a company were classified into two categories: those held continuously since the previous period and those purchased during the current period. Those stocks purchased in the past were valued at prices based on the previous period, while those recently purchased were valued at current market prices. The ordinance aimed to prevent investors from being penalized from the recent stock price fluctuations, which were attributed to the recent international situation and not to the soundness and performance of the company itself.

Reform bureaucrats claimed that the government was in no way biased against equity and recognized it as "one extremely powerful method of obtaining funds for expansion of production" as well as "an important means for national savings." In view of the policies of the Financial New Order, however, it was clear that bureaucrats encouraged indirect over direct financing and long-term over short-term investment. Price controls, dividend controls, and control over management via the creation of a new industrial order effectively dampened enthusiasm for stocks. Moreover, the government's savings campaign, which targeted white collar and blue collar workers, tended to channel funds to banks rather than to the stock market.[59]

The New Order for Science-Technology

Science and technology played a central role in the New Order. Reform bureaucrats believed that the key to Japan's future lay in developing a new form of Japanese science-technology based on the resources of Asia. In the spring of 1940, the movement for a New Order for Science-Technology (*Kagaku-gijutsu shintaisei*) was launched by reform bureaucrats and technical experts at the Cabinet Planning Board's Science Division and the Asia Development Board's Technology Section. The purpose of the movement was to make scientific and technological research a national priority, promote experts of science and technology to

---

59. Okazaki, "The Japanese Firm under the Wartime Planned Economy," 365; Okazaki, "Wartime Financial Reforms and the Transformation of the Japanese Financial System," 145.

policymaking roles, and reorient scientific and technological research from a liberal capitalist approach to a state-centered, nationalist approach.

In the Cabinet Planning Board's "Outline for the Establishment of the New Order for Science-Technology" of May 1941 bureaucrats pointed out the problems with Japan's system. They bemoaned the general lack of interest in science and technology, particularly among Japan's ruling class. The common view among politicians, scholars, and educationalists was that science and technology were matters that should be left to scientists and engineers. Business leaders tended to opt for the cheaper and quicker route of purchasing foreign technical patents rather than developing technology in-house. In addition, they criticized the poor treatment of technical experts within the public and private sectors. Technical officials (*gikan*) in the bureaucracy had a lower rank than administrative officials (*jimukan*) and were largely excluded from policymaking, while most technical specialists were denied promotion to senior posts. Not surprising, many university students preferred to major in law or literature instead of science. The general lack of commitment of Japan's leaders to science and technology resulted in Japan's diminished competitiveness abroad. The Cabinet Planning Board pointed out that although top-rate scientific research was being carried out at the university, its findings either remained within academia or were sold abroad rather than being applied for the benefit of the Japanese state. Officials bitterly recalled how Japanese leaders had sold cutting-edge research on synthetic petroleum to Germany, which quickly applied it to military industrial production with spectacular results. In the private sector, imported and domestically developed advanced technology was treated as a proprietary good that was monopolized by firms in order to enhance their competitiveness vis-à-vis other firms.

As the first step in making technology policy a national priority, bureaucrats created a new cabinet-level research and policymaking body, the Technology Board (Gijutsuin). Miyamoto Takenosuke at the Asia Development Board in May 1940 drafted plans for the new organization. Modeled on Ōkōchi Masatoshi's Continental Science Board in Manchuria, the Technology Board was placed directly under the office of the prime minister. Its mandate was to centralize all science and technology research and policymaking functions of the various ministries in order to "comprehensively display the state's total powers of science-technology and aim for its reform and advance, particularly in aircraft-related science-technology."[60] The establishment of the Technology Board was

---

60. Nihon kagakushi gakkai, ed., *Nihon kagaku-gijutsushi taikei*, vol. 4 (Tokyo: Daiichi hōki shuppan, 1966), 357.

delayed, however, because the various ministries refused to yield their technically related areas of administration to it. The army and navy insisted that the board focus specifically on aircraft technology for the war instead of taking a broader, long-term approach toward science and technology policy.[61] The Technology Board was finally established in January of 1942 but in heavily compromised form and its main national policymaking role was taken over by the more effective Science-Technology Council, founded in December of 1942.[62]

The Technology Board provided scientists and engineers with a new status. It was staffed predominantly by technical bureaucrats, who were appointed "technical advisors" (*sangikan*) with the equivalent rank of administrative officials and given policymaking roles.[63] The creation of a new rank for technical officials represented a major reform of the Civil Service Appointment Ordinance. Following the bureaucratic reforms of November 1942 the Ministry of Railways referred to technical and administrative bureaucrats as "railway bureaucrats" charged with both administrative and technical duties.[64] Such a blurring of the roles of technical and administrative bureaucrats had been promoted by both Mōri and Miyamoto. Miyamoto called on engineers to change themselves from narrow specialists to "administrative engineers" (*Verwaltungsingenieur*) who "maintain contact with all fields including government, economics, and culture, and display the synthesized results."[65]

Technology Board officials sought to liberate technology from the profit-driven economy and place it in the service of the state. Officials put forth a "Five Year Plan for Technological Advance," which targeted specific areas of research considered vital to national interests. Scientific and technological research was interpreted as a public good and its import and export were controlled. In order to eliminate the monopolization of industrial property rights or patents, which officials viewed as a "relic of capitalism," the state applied restrictions on its use and expropriation and promised to compensate firms accordingly. In addition, they sought to incorporate more technical experts into the management structure

61. Sawai Minoru, "Policies for the Promotion of Science and Technology in Wartime Japan," *Keizaigaku ronshū* 35, no. 1 (June 1995), 59–60.

62. Hiromi Mizuno, *Science for the Empire: Scientific Nationalism in Modern Japan* (Stanford: Stanford University Press, 2008), 67–68.

63. Ōyodo Shōichi, *Gijutsu kanryō no seiji sankaku* (Tokyo: Chūō shinsho, 1997), 182. On the Continental Science Board, see Kawahara, *Shōwa seiji shisō kenkyū*, 84–92; Mimura, "Technocratic Visions of Empire: Technology Bureaucrats and the 'New Order for Science-Technology,'" 107–110.

64. Ōyodo, *Gijutsu kanryō no seiji sankaku*, 183.

65. Miyamoto Takenosuke, "Kōa gijutsu no konpon genri," in Miyamoto, *Tairiku no keizai kensetsu*, 152.

of firms and to reward firms that promoted technical personnel and contributed to the rapid advance and dissemination of superior technology.

The most difficult challenge was to change the spirit of technology and develop Japan's own indigenous form of science-technology. Officials sought to change the image of technology from a narrow, technical field subordinated to universal science to a national asset grounded in the Japanese ethnic community and the resources of Asia. The term "Japanese science-technology" came into vogue around the time of the launching of the New Order and appeared in the Cabinet Planning Board's outline, which proclaimed the anticipation of "the completion of a Japanese character of science-technology based on the resources of the Greater East Asia Co-Prosperity Sphere." It was taken up in debates about the spirit and role of technology among reform bureaucrats, technical bureaucrats, and left-wing intellectuals such as Aikawa Shūji, Aikawa Haruki, and Miki Kiyoshi. The term was adopted to clarify the close relationship between science and technology and highlight technology's important, strategic character. Whereas in the past technology had been viewed as a spin-off of science, now it was recognized as a separate discipline. As Aikawa Haruki pointed out, science was extremely contemplative, cognitive, and sought to reveal natural laws. In contrast, technology was directly aimed at action, practice, and the application of natural laws and possessed its own unique, multifaceted character shaped by its intimate relationship to production, labor, and scientific culture.[66] Cabinet Planning Board officials argued that in order for Japan to establish an indigenous form of science-technology, it needed to pursue its own scientific and technological research based on the resources of the Greater East Asia Co-Prosperity Sphere and direct it toward achieving national objectives. They noted that a technology that utilized the resources of England was of no value to Japan in this day and age.[67] Mōri called for the development of a "political character of technology" in which technology was directly tied to the spiritual power of the people.

Although vague about the precise methods in which these goals would be achieved, planners made clear that such reforms would be possible only after a New Order for Industry was established. Aikawa Shūji suggested that technology could be liberated only when management as a whole was liberated from profit-based capitalism and a management cooperative body was established.[68] Mōri believed that the process of technology's spiritual transformation began at the top. As he explained, the revolution of the spirit of technology was possible only through a revolution of its organization, which took as a precedent the

---

66. Aikawa Haruki, "Shintaisei to gijutsu no soshikika," *Gijutsu hyōron* (January 1941), 187.
67. Kikakuin kenkyūkai, *Kokubō kokka no kōryō*, 187.
68. Aikawa Shūji, "Gijutsu no kaihō to rōdō no rinen," Minobe Yōji Documents, L 1:80.

revolution of the organization and spirit of enterprise, which in turn was effected by means of the establishment and systematization of the spirit of the national economy.[69]

The New Order for Labor

The "liberation" of management and technology from capital had profound implications for labor. Under the liberal capitalist system, labor's relationship to both management and technologists was class-based and antagonistic. Management, which identified with shareholder interests, viewed labor as a means of production whose costs should be minimized to increase profits. Technical experts also identified themselves with the interests of capital. The left-wing reformist Aikawa Shūji criticized the "egoism" of technologists, especially their tendency to view labor as a "slave of the machine" and a "wage proletariat," whose material labor power was a commodity that could be bought and sold cheaply.[70]

Through the establishment of a New Order for Labor, reform bureaucrats, labor leaders, and left-wing intellectuals attempted to replace this antagonistic, class-based relationship with one in which management, technology, and labor were organically fused together into a harmonious, functional management body focused on productive power rather than productive relations.[71] As the Cabinet Planning Board explained, by separating capital from management, the relationship between management and labor would evolve into one of close cooperation based on equally respected but functionally differentiated roles.[72] Moreover, they called for the establishment of a new spirit of labor in which work was viewed as "honorable" (*kinrō*) and workers become conscious of themselves as members of the "country of our emperor" (*kōkokumin*) whose labor represents dedicated service to the state.[73] Some reformists on the left also hinted at the promises of higher wages and better employment terms. In their proposals for economic reorganization, the Shōwa Research Association painted a rosy picture for workers by suggesting that the fulfillment of labor's functional roles within the firm would be rewarded with an "active expansion" of the compensation system in addition to the honor of working under publicly recognized management.[74]

---

69. Kamakura Ichirō, "Tōa kyōdōtai to gijutsu no kakumei," *Kaibō jidai* (March 1939), 11.

70. Aikawa, "Gijutsu no kaihō to rōdō no rinen."

71. "Shintaisei kensetsu yōkō," Minobe Yōji Document, G 2:14; Tanaka Shinichi, "Sangyō ni okeru 'shintaisei' an yōkō (sōan)," Minobe Yōji Documents, G 2:4.

72. Kikakuin kenkyūkai, *Kokubō kokka no kōryō*, 213.

73. Saguchi Kazurō, "The Historical Significance of the Industrial Patriotic Association: Labor Relations in the Total-War State," in Yamanouchi, Koschmann, and Narita, eds., *Total War and 'Modernization,'* 271–272.

74. Shōwa kenkyūkai, "Nihon keizai saihensei shian," 18–19.

Officials claimed that the movement for a New Order for Labor represented a new stage in labor policy. Labor policy had evolved from a liberal labor policy to one based on the Industrial Patriotic Association (Sanpō) and was now entering a new era. They justified the Labor New Order on two grounds. First, the changing international situation called for a new policy toward labor management. Japan's increasing isolation from the world's resources compelled Japan to change its export-oriented industrial policy to one aimed at the establishment of a national defense state system, the expansion of productive power, the rationing of resources, including labor, and the turn inward toward a self-reliant sphere. Second, planners were concerned about the dismal state of the workplace in the late 1930s. To address the problem of labor shortages, low productivity, and frequent job switching, which was said to occur among 81 percent of workers, the government passed a series of labor control ordinances based on the National General Mobilization Law.[75] In 1939 ordinances regulating wages, hours, the hiring of school graduates and youth, and job transfers were passed. Controls on job changes were further strengthened and culminated in 1941 in the institution of a Nazi-style employee passbook system to provide a record of a worker's employment history to future employers. Despite increased government controls, the problems of labor shortages and low productivity, particularly in the chemicals, machine, and metals industries, remained acute.

Cabinet Planning Board planners acknowledged that worker dissatisfaction lay at the heart of the problem. In their plan for the "Establishment of a New Order for Labor" of November 1940, they made it a key policy to raise the status of workers and to inculcate a spirit of labor within the nation. In their commentary on this plan, they praised the Nazi labor policy because it gave labor a central place within its "ethnic cooperative body" (*Volksgemeinschaft*). Since the late 1930s, Nazi policies to promote a classless, organic body composed of capital and labor had become popular among reformists. The Home Ministry bureaucrat Minami Iwao was an enthusiastic supporter of the Nazi-style discussion councils composed of labor and capital. He praised Nazi Germany's "strength through joy" program, which sought to motivate workers by providing entertainment and travel packages. At the Cabinet Research Bureau, Minami drafted a plan for a Nazi-type labor program, which served as a model for the Industrial Patriotic movement. Kamei Kan'ichirō supported the German Labor Front's call for the creation of a new "production class," composed of technology and management experts and labor, and based on the idea that "whether capitalists, engineers, or laborers, they represent differences in vocation within the complex state

---

75. Kikakuin kenkyūkai, *Kokubō kokka no kōryō*, 203.

industry, not classes."[76] Like Minami, Kamei was attracted to the Nazi's "strength through joy" movement because it raised the social position of workers and enhanced their pleasure. Aikawa Haruki suggested that the "management cooperative body" ought to be termed a "labor cooperative body" because managers, technologists, white collar workers, and blue collar workers were all essentially laborers; the only difference was that they possess different functions.[77]

Reformists' claimed that their promotion of labor went beyond mere rhetoric. They proposed the creation within each management body of a special corporate organization (*tokubetsu shadan soshiki*) comprised of managers and workers. The firm's productive character would be displayed by its employees to the highest degree through the spirit of cooperation based on their respective functions. In addition, organizations would be established to improve productivity and advance the welfare of workers through a two-way dialogue between managers and workers. Following the establishment of the Great Japan Industrial Patriotic Association in November 1940, reformists incorporated these ideas into the association's business plan. Among the items were the establishment of discussion councils (*kondankai*) composed of owners and workers, research bureaus to investigate industrial labor issues, and measures to advance workers' culture along the lines of the Nazi's "strength through joy" movement. The cultural program included the sponsoring of theatrical companies, factory bands, international labor cultural exchange, and exhibitions of workers' products.[78]

Labor organizations in wartime Japan did not enjoy the prominent place and identity as in Nazi Germany. The Nazi structure was based on the three separate pillars of the industrial companies, Labor Front, and the Nazi government. Although workers were obligated to serve the state, their separate identity and base of power was acknowledged. In Japan, the various components were portrayed as forming one family based on Japan's ancient family system. The head of the control association represented the family head, which was responsible for providing for the well-being of the workers under its care. As Minobe explained, "In terms of priority of public interests, Japan's ancient spirit lies in the idea of unity between public and private—the individual lives together with the whole, the whole nurtures the individual within. This is the ancient spirit of Japan."[79] Workers had no separate identity in this system but were viewed as possessing equal status vis-à-vis other groups. They would be treated as members of one

---

76. Kamei, "Doitsu no sangyō seishin," *Kagakushugi kōgyō* (August 1938), 155, 158.

77. Aikawa Haruki, "Shintaisei to gijutsu no soshikika," *Gijutsu hyōron* (January 1941), 13–18.

78. Alf Lüdtke, "The Honor of Labor: Industrial workers and the Power of Symbols under National Socialism," in David J. Crew, ed., *Nazism and German Society* (London: Routledge, 1994), 66–109.

79. Minobe Yōji, "Tōseikai no riron to shōrai mondō," in Minobe, *Senji keizai taisei kōwa*.

large family instead of as simply a means of production and the property of employers.

Defending the New Order

The controversy over the New Order revealed a changed political landscape. Among the most vocal critics were politically powerful industrialists such as Commerce Minister Kobayashi Ichizō, Railway Minister Ogawa Heikichi, Communications Minister Murata Shōzō, and Home Minister Hiranuma Kiichirō, the doyen of the idealist right wing. They defended the traditional prerogatives of big business and the bureaucracy and denounced the New Order as a communist plot. The new coalition of business leaders and the right wing against a coalition of reformists and leftists revealed how much the terms of political debate had changed in wartime Japan.[80] Although the reform movement had first joined with the radical right to attack capitalism in the early 1930s, it gradually allied with the left in promoting state planning and a new technological worldview. Conservatives, who had joined with the progressives in advocating a more pluralist system, now found common cause with the right in defending the Meiji constitutional order and opposing state control.

The heated debates over the Economic New Order revealed the high stakes involved. Those aligned with the forces of reform such as military and civilian technocrats, technical experts and specialists such as engineers, scientists, new zaibatsu leaders, economic planners, and employees of the advanced industries could expect a bright future in the New Order. Those identified with the status quo such as former party politicians, zaibatsu leaders, small- and medium-sized business owners, traditional bureaucrats, and landlords were compelled to adapt to the rapidly changing political economy or face redundancy.

Separating Capital and Management

The distance between reformist and conservative positions was made painfully clear in the debates over "separating capital and management." The views of business were conveyed in numerous policy statements such as the "Opinion Concerning the Economic New Order" of November 1940 put forth by the Japan

---

80. Itō, "The Role of Right-Wing Organizations in Japan," 487–491; Kinmonth, "The Mouse that Roared," 331–332.

Economic League and six other business organizations.[81] Business pointedly questioned the wisdom of launching a radical movement for reform when the fate of the nation was at stake in the China War. In their drive to correct the evils of the so-called "liberal economy," they argued, these reformists shook the foundations of the economy, created social unease, and demoralized the people at a time when everyone should be devoting their energies to expanding productive power at the highest efficiency levels for the purpose of building the national defense state.

Business upheld the profit motive. They criticized bureaucrats for failing to understand the true nature of business and the firm. The pursuit of profits or fusion of capital and management, they argued, was an essential feature of the economy. It was the key element in increasing productivity, obtaining valuable imports, and financing the national defense state. By rejecting the profit mentality in favor of lofty bureaucratic principles such as "public over private," "selfless patriotic service to the state," or "production over profits," reformists jeopardized the firm's performance. Ironically, they praised the Nazis for recognizing both the legitimate role of "pure profits" in business and the close relationship between capital and management.

Business also appealed to the rightist cause by exalting the Japanese spirit and its sacred national essence, or *kokutai*. They accused reformists of being too westernized. By closely identifying the Japanese economy with the problems of Western liberalism, reformists failed to recognize the native roots of business's creativity, talent, vitality, and courage. They warned that by identifying the pursuit of profits with the Japanese spirit, reformists threatened to destroy the foundations of Japan's economy and bring about a Russian-style revolution. Moreover, business denounced bureaucratic intervention in the activity of the firm, particularly with regard to policies to merge firms, rationalize industry, and control profits and prices. They cautioned that the "one company, one industry" policy would destroy the multilateral holding companies and that the ordinances to control company finances and accounting, especially dividends and remuneration, contradicted the spirit of the firm and the expectations of workers and management.

As chief spokesman for the reformist camp, Minobe took the lead in presenting a formal rebuttal to business's policy statements. In typical reformist fashion, he justified the state's policies from the standpoint of the deductive goals of the state and the logical connection between ends and means. The overall goal of the Economic New Order, he began, was to establish the national defense

---

81. Minobe Yōji, "Keizai shintaisei kanken," *Nihon hyōron* (March 1941), 12–20; Nakamura and Hara, "Keizai shintaisei," esp. 96–97.

state system. In order to realize the national defense economy the state needed to increase production of munitions, increase productive capacity to secure the necessary materials for munitions production, and ensure the basic necessities for daily life. These objectives could only be achieved if the country's limited supplies of materials, labor, and funds were utilized in the most efficient manner through prioritization of production and the management of the national economy in a planned manner.

Minobe criticized business for failing to acknowledge both the fundamental challenges facing the nation and the very limited options available given the international situation. He insisted that amidst these challenging times the government was not pursuing "reform for the sake of reform," but rather for the purpose of establishing the national economy on a planned basis and expanding production by maximizing the efficiency of total state power.[82] Minobe emphasized that reforms were not being pursued for ideological reasons—in other words, to promote Marxism. Reforms arose not from a desire to correct the evils of the so-called liberalistic economy or because its defects called forth the need for reforms. Rather, reforms arose from the need to comprehensively manage the national economy for the purposes of war. In other words, Minobe argued, reform was not the end but the means and the means were not Marxist but Japanist and founded on Japan's native *kokutai*. With regard to the issue of profits, he asserted that profit itself was not rejected. However, there was no private profit without public profit. He advocated the fusion of public and private profits and the "purification" of profits.

Minobe noted, on another occasion, that from the theoretical standpoint, a planned economy had no necessary connection to capitalism or socialism; it was simply "an economy that takes the unified will of the state as its subject." The central issue in planning was productive power not productive relations. As imperial bureaucrats and implementers of the "state's will," they sought to increase production, expand productive power, and provide for the basic necessities of civilian life within the constraints of limited resources. Only through planning could they efficiently utilize and channel scarce resources to priority industries and reinvest capital into the firm. Moreover, bureaucrats justified their policies on the grounds that planning was the inevitable historical trend in all modern industrial countries. As a result of the shortcomings of the liberal market system, all nations sought to intervene in the economy. Politics took precedence over economics and gave rise to economic planning in which the invisible hand of the market was replaced by the visible hand of the state, collective interests overrode

82. Minobe Yōji, "Keizai shintaisei kanken," 18.

individual interests, and production not profit became the driving force. When the economy was no longer driven by profits, management would become "liberated" from the control of capital and capital and management would be separated.

Reform bureaucrats vehemently denied charges that they were promoting socialism and emphasized their close links to the idealist right and Nazism. Sakomizu regretfully noted that people automatically assume that terms such as *managed state* and *planned economy* are Russian terms implying the rejection of private property.[83] Mōri emphasized the German roots of planning. He credited the former German economic minister Hjalmar Schacht for being a driving force behind the planned economy. The reformist policy of priority of public interests was adapted from the German concept of "public interests over private interests" (*Gemeinnutz vor Eigennutz*). Reformists insisted that they rejected neither profits nor private property. Their point of departure for planning was neither the socialist view of class conflict nor the communist ideal of a classless society but rather "each and every citizen assisting in the imperial rule of an entire nation cooperating together as one hundred hearts beating as one under the (new) order."[84]

In the final draft of December 1941, reform bureaucrats agreed to permit business to choose their leaders. They also reduced the antiliberal tone of the draft, but at the same time retained the key concepts of the advanced national defense state, leadership principle, and planned economy. Reference to the controversial term *separating capital and management* was replaced with the less antagonistic concept of "fusing capital, management, and labor."

Fusing Public and Private

Business leaders were disturbed by not only the contents of the proposals, but by the unilateral and secretive manner in which they were drafted and then "dropped" like a bomb on the public. They complained that the manner in which reform bureaucrats conducted policymaking was, if anything, promoting bureaucratic-civilian estrangement not cooperation. Business denounced the Cabinet Planning Board Deliberation Room staff as political upstarts and self-appointed *rōnin* who circumvented official bureaucratic channels to gather together and secretly draft national policy in an office that was not even officially on the books.[85] They

---

83. Kashiwara, Minobe, et al., "Zadankai: Kakushin kanryō—shintaisei o kataru," 64.

84. Ibid., 66.

85. Kitagawa Kazuo, "Zaikai to kakushin kanryō," *Kaizō* (January 1941), 156–157; Shimada Hirosaku, "Kishi Nobusuke ron," *Kaizō* (December 1941), 224.

argued that bureaucrats were government employees and should stick to their officially sanctioned role as civil servants and stay out of politics.

At the Emergency Central Cooperative Conference, business leaders advanced three proposals: experts from the private sector assume control over the Cabinet Planning Board's Deliberation Room and draft the economic plans; civilians should be assigned to the various ministries as economic attachés; and civilians participate in and obtain training at the new Total War Research Institute. In subsequent negotiations with Cabinet Planning Board leaders, business tried to secure a commitment to staff half of the Deliberation Room with private civilians. Although it appears that business supported the idea of public and private collaboration, they did so for pragmatic reasons. For business, collaboration reflected a desperate attempt to retain some influence in the new state-controlled system. As one executive put it, business would become "dependent directors of a beggar controlled economy" in which industrialists were forced to work for the grand prize of remuneration and lose their independent spirit; they might be released from the shackles of capital but now they would be chained to the bureaucracy.[86]

Critics sensed the far-reaching implications of the New Order and the aim to fundamentally change the basic structures of power in business and government. In one roundtable, reform bureaucrats bluntly stated that whereas in the past, business and the bureaucracy had operated in separate worlds, now those lines would be blurred and traditional leaders in each sphere would be replaced by a new breed of technocratic leaders, or führers. In the same way that bureaucrats must change from legal bureaucrats to creative bureaucrats, business leaders must change from representatives of capital to public servants who possess a sense of public responsibility and the technocratic skills and vision of a führer. Mōri defined the führer as one who calmly led the masses, possessed a strong conviction, and considered the needs of the whole; among the qualifications was "the ability to bring together a total grasp of the dynamic reality and the deductive goals of the state and assess them in comprehensive fashion."[87]

The political implications of the New Order were disturbing for members of the status quo, particularly those in senior administrative and management positions. For the traditional bureaucrat trained in law and uncomfortable with the new industrial and management technologies, the New Order spelled the end of their careers. Mōri warned that those bureaucrats who did not possess the skills of the führer would not be considered true bureaucrats and should resign

---

86. Kitagawa, "Zaikai to kakushin kanryō," 157–158.
87. Kashiwara, Minobe, et al., "Zadankai: Kakushin kanryō—shintaisei o kataru," 54.

from their posts.[88] Both business and bureaucrats viewed the proliferation of technocratic planning bureaus within the bureaucracy with alarm. Now reformists were hinting at the need for a "bureaucratic new order" and the creation of a leadership organization within the administrative structure itself. As Minobe suggested, the Economic New Order would not be possible without the creation of a suitable bureaucratic structure to operate the planned economy.

What was the New Order? Was it a socialist plot to launch a "red" revolution and thereby crush the power of capitalists and overthrow the emperor? The answer varies depending on one's political position. From the standpoint of communists, who rejected the emperor system, the reform bureaucrats and everyone else were right-wing. Socialists and leftists, in contrast, believed that one could reject capital and accept the emperor system. Some viewed reform bureaucrats as a progressive force because of their anticapitalist stance, technological vision, and elevation of labor's status within a functionalist society. From the perspectives of old-school bureaucrats, businessmen, and military officers, as well as imperial advisors and the idealist right, the reform bureaucrats were "red." The "red" label was a convenient term that identified the imperial institution with capital and landownership. Hence, the principle of "separation of capital and management" was perceived as a rejection of both capital and the emperor. This chapter suggests that the New Order was neither right nor left. Its techno-fascist vision sought to transcend both and overcome the conflicts of interest and tensions among the various groups.

---

88. Ibid., 54–55

6

JAPAN'S OPPORTUNITY
Technocratic Strategies for War and Empire,
1941–45

In 1915 Thorstein Veblen prophetically wrote about a temporary window of opportunity for Japan to combine its national spirit and recently acquired industrial technology with maximum effect in a major military offensive. Veblen predicted that the window would gradually close as modern technical advances eroded traditional notions of community and loyalty and introduced a materialistic and commercial mindset bringing about the "sabotage of capitalism."[1] A quarter of a century later, however, Japanese technocrats remained exceedingly optimistic about their country's prospects for war and empire. They were determined to "overcome the modern," despite the attempts of conservative businessmen and bureaucrats to derail the New Order movement. Technocrats believed that, more than Japan's material power, its human resources, namely the patriotic spirit, courage, discipline, and creativity of its people were the fount of national strength. Spiritual mobilization, they argued, was just as important as material mobilization for war. Mōri Hideoto was confident that Japan would excel in both. He claimed that Japan was one of the few nations that had in place an advanced national defense state that regulated every aspect of society from the standpoint of defense and was mobilized for war.[2]

---

1. Thorstein Veblen, "The Opportunity of Japan," in *Essays in Our Changing Order* (New York: Viking Press, 1952).
2. Mōri Hideoto, "Kōdō kokubō kokka no tatsu" (November 1940), Mōri Hideoto Documents 2:7.

On August 1, 1940, Foreign Minister Matsuoka Yōsuke announced the government's policy to build a so-called "Greater East Asia Co-Prosperity Sphere." The term *Greater East Asia* implied that in addition to the core region of Japan, Manchukuo, and China, the sphere would include Southeast Asia, Eastern Siberia, and possibly the outer regions of Australia, India, and the Pacific Islands. The new policy to expand the boundaries of Japan's empire beyond East Asia emerged after France and the Netherlands fell to Nazi Germany in the late spring of 1940 and forfeited their colonies in Southeast Asia. Japan subsequently advanced into French Indochina in June 1940. Three months later in September 1940, Japan concluded the Triple-Axis Pact with Germany and Italy. When diplomacy failed to lift economic sanctions imposed on Japan by the United States, Japan attacked Pearl Harbor on December 8, 1941. These actions set the country on a course of brutal occupation of Asia and a destructive war against the United States and its allies that culminated in Japan's total defeat in 1945.

The question of why resource-poor Japan would take on the world's superpower continues to baffle analysts of the wartime period. Statistical estimates of declining national income, industrial production, foreign currency reserves, and a growing trade deficit, accompanied by a steady stream of state control laws and ordinances, suggested the reality of decreasing national wealth and power and desperate attempts to reverse the decline. It is well known that between 1937 and 1945 the Japanese state squeezed the economy through strict rationing in the civilian sector and the control of management and labor in order to channel a dwindling supply of precious resources to the military's ambitious production expansion and material mobilization plans.[3] The drain on resources from the protracted war in China, food and energy shortages, higher import costs as a result of the European war, and rapidly deteriorating trade relations suggested that Japan had little chance of victory in a war against the United States. Japan's leaders grossly overestimated their country's economic and military capacity and underestimated that of its opponents. Their decision to bomb Pearl Harbor was certainly a consequence of economic and political miscalculation, short-sightedness, ignorance, and overconfidence.

Japan's decision to start the war can be better understood by examining the ideology and politics of its leaders. Technocrats conceived of the impending war as more than a battle of resources. They saw it as a battle of worldviews that was being fought in Asia and the Pacific against the Western liberal powers. They were convinced that the capitalist world order was being surpassed

---

3. Hara Akira, "Japan: Guns before Rice," in Mark Harrison, ed., *The Economics of World War II: Six Great Powers in International Comparison* (Cambridge: Cambridge University Press, 1998); Cohen, *Japan's Economy in War and Reconstruction.*

by a new type of geopolitical world order based on technocratic planning in which Japan would have a prominent place.[4] The key issue concerned the nature of planning—should it be liberal, fascist, or communist? Technocrats viewed liberal-capitalist planning as extremely inefficient and ultimately doomed. Communism was perceived as despotic and antithetical to Japan's imperial system. Fascism represented a third way and the most promising planning route to power because it combined advanced technology with national spirit. Fascist planning was the preferred mode of planning for "have-not" countries with great power aspirations such as Japan and Germany. It represented an attempt to compensate for a lack of material resources with Japan's spiritual resources and the power of science-technology. From the standpoint of military technocrats, the Pearl Harbor attack was not only a wager to force the United States to accept Japanese hegemony in Asia, but also a means of reform. They viewed the Pacific War as the first step toward constructing a technologically advanced, self-sufficient, regional economic sphere, or *Grossraumwirtschaft*. Reflecting the reformist view of war as an integral part of state reform, the new Cabinet Planning Board Chief Akinaga Tsukizō proclaimed that Japan would "build while fighting" (*tatakainagara kensetsu e*).[5]

From the perspective of civilian technocrats, the Pacific War provided an opportunity to complete the unfinished business of the New Order. In 1941 a new constellation of forces emerged. The nexus of planning shifted from the Cabinet Planning Board to the Commerce Ministry, and later to the Ministry of Munitions. Kishi changed his strategy to fostering closer relations with business and reforming the bureaucracy. Although big business had preserved many of its privileges within the new economic system, it lost its autonomy in the face of the pressing demands of total war. Shiina wryly remarked that the Pacific War represented the "baptism" of business by which industrialists were finally compelled to embrace war mobilization and become partners of the state.[6] As a result of the diminished authority of the Cabinet Planning Board, the state's roundup of the left, and the dissolution of technocratic organizations such as the Shōwa Research Association and Social Masses Party, the left lost its institutional base of power and political influence. The traditional right wing, in contrast, had powerful spokesmen in the military, bureaucracy, business, and the Diet and flourished within the new wartime regime.

---

4. Minobe Yōji, "Shin keizai taisei no kōsō," *Jitsguyō no sekai* (November 1941), 26; Kishi Nobusuke, Akinaga Tsukizō et al., "Sangyō shintaisei no shinro," *Kagakushugi kōgyō* (April 1941), 106.

5. Akinaga Tsukizō, "Idai naru kokuryoku no saikentō," *Jitsguyō no sekai* (October 1941), 21.

6. Shiina Etsusaburō, "Nanpō shinshutu no kamae," *Jitsugyō no Nihon* (March 1942), 20–25.

Completing the Advanced National Defense State

In the early months of 1941, reformists expressed concern about the public's dampened enthusiasm for the New Order movement. In one roundtable, Akinaga recalled how certain politicians at the recent Diet meeting announced their strong reservations about the Economic New Order and seemed to be organizing activities to discredit it.[7] It appeared that the movement had reached a crossroads in which it could either flourish and develop along reformist lines or stagnate and congeal into the liberal "status quo" mold. Reformists felt it crucial to sustain the momentum of the advanced national defense state and preserve the spiritual unity of the nation. To allow the movement to stagnate would not only jeopardize the decade-long efforts of reformists but place Japan in a critical predicament since it was becoming increasingly cut off from outside resources. The domestic challenges faced by leaders were threefold: to obtain the cooperation and expertise of business; to overcome bureaucratic sectionalism; and to convince the skeptics of Japan's preparedness for war.

Lessons of the Economic New Order

Despite the different circumstances in Manchukuo and Japan, reformists faced the similar challenges of gaining the support of business leaders and devising ways to effectively combine government planning with private initiative. In both cases, a pattern can be seen in which reformists alienated business early on by brandishing ideological slogans such as "zaibatsu stay out" and "separating management and capital" and subsequently extended an olive branch by clarifying and recasting their policies in more business-friendly terms. This pattern reflected the ideological diversity and functional division of roles among members within the reform faction. The visionaries and theorists tended to be farther to the right or left of those pragmatic, centrist reformists in charge of implementation. In Manchuria, Kishi's group had devoted considerable energies to recasting and modifying the vision and economic policies of the Kwantung Army, left-wing economists, and pan-Asianists in order to present them in a more favorable light to business and the government. Now in Japan, the theories and concepts devised by Akinaga and his planning triumvirate at the Cabinet Planning Board were modified by the more business-minded bureaucrats at the Commerce Ministry. Following the uproar over the idea of "separating capital and management," reform bureaucrats tried first to emphasize the fascist influences on planning

---

7. Nobusuke, Akinaga, et al., "Sangyō shintaisei no shinro," 101.

and later denied the ideological aspects altogether. The latter strategy proved to be more effective in rallying support from business and the public.

Reformists acknowledged that skillful marketing of the idea of state control was as important as the policies themselves. In a press interview in August 1942, Kishi drew a clear distinction between the new control measures and those of the Konoe era. He complained that there was "too much theory" in the proposed reforms and stressed that implementation, not theory, was the overriding concern.[8] Kishi's new "nonideological" approach was just as ideological, however. It expressed the consciousness of the technocratic elite, which claimed neutrality, impartiality, and freedom from class interests while promoting the authority of the expert and professional class.

The other conclusion that reformists drew from their recent planning attempts was that bureaucratic control did not work and that the state should defer to business leaders on issues concerning the internal management of their firms. In Manchuria reform bureaucrats had abandoned the special company system based on the principle of "one industry, one company" and turned to Ayukawa to reorganize and consolidate the special companies under Nissan's own corporate structure. In Japan, these bureaucrats were coming to similar conclusions in light of the lackluster performance of the twenty-two industrial control associations. The control associations were heavily criticized for being powerless, ineffective, and failing to meet the government's production targets.

In a scathing report on the control associations, the Cabinet Planning Board identified the sources of their weakness.[9] The first problem was the lack of enthusiasm and support from business. Although the control associations were nonprofit entities, the member firms continued to operate on a profit basis. The report accused business of sabotaging the control associations by refusing to supply its best managers, denying control association inspectors access to factories, and generally obstructing the smooth functioning of the control associations. It exhorted business to correct their attitude and actively cooperate with the control associations.

The second problem was the heavily bureaucratic character of the control associations. The report emphasized that the control associations were not meant to be bureaucratic organizations, but rather semiprivate, semipublic entities that nurtured a close relationship with the private sector. The report accused the control associations of forgetting their original mission and becoming no more than an additional administrative layer, rigid and unresponsive to the needs of the

---

8. Kishi Nobusuke and Noyori Hideichi, "Senjika no keizai seisakutaidan, Part 1," *Jitsugyō no sekai* (August 1942), 53.

9. Kikakuin kenkyūkai, *Tōseikai no honshitsu to kinō* (Tokyo: Dōmei tsūshinsha, 1943), 28–33.

member firms. The Cabinet Planning Board recommended that the control associations develop their own research capabilities and establish a Board of Trustees to keep abreast of new developments within its industry. In addition, it criticized the control association leaders for insufficiently embracing their public mission and using their positions to obtain information for their private firms.

The third problem was the lukewarm, noncommittal attitude of the bureaucracy. The Cabinet Planning Board pointed out that the control associations were not just the responsibility of the Commerce Ministry, but also the ministries of finance, welfare, and agriculture and forestry. These ministries needed to overcome their sectionalism and completely transfer the relevant areas of jurisdiction to the control associations. Finally, the report also pointed out various defects within the structure of the control associations, including the leader's insufficient authority with regard to personnel matters and the lack of coordination of production and distribution among member firms.

Reform bureaucrats began to explore alternatives to the control associations. When asked about the need to strengthen the control associations in September 1942, Kishi replied: "With regard to the control associations, this might sound like an extremely presumptuous claim, but I do not seriously think that we need to greatly strengthen them."[10] Kishi was more enthusiastic about the new public management foundations (*eidan*) and financial depositories (*kinko*). These nonprofit organs were set up to manage and finance projects considered financially unprofitable but vital to the public, including worker housing; conversion of idle factories; Southeast Asian development; and loans to agriculture, small business, and company employees. The real challenge for reformists was to find a way to obtain the cooperation and expertise of top-flight business leaders. Given the top-down authoritarian nature of the control associations based on the so-called führer principle, the fate of the control association was completely dependent on the ability of the leader to effectively manage the member firms and command their respect and allegiance. Bureaucrats were unable to recruit capable business leaders who could raise the stature of the control organizations both in the eyes of business leaders as well as bureaucrats.

In a major shift in strategy, Kishi struck a compromise with business in the form of the new Munitions Corporation Law in October 1943. Similar to the arrangement made with Nissan in Manchuria, Kishi enticed certain companies to expand production in munitions-related areas and meet government targets by providing state subsidies and financial guarantees. The new law essentially empowered the government to bypass the control associations and work directly

---

10. Kishi Nobusuke and Noyori Hidekichi, "Senjika no keizai: Seisaku taidan, Part 2" *Jitsugyō no sekai* (September 1942), 71.

with selected munitions firms to achieve production targets. Under the law, the government granted certain firms official status as "munitions companies." In order to enhance their status, the principle of "separation of capital and management" was upheld. In each company, a responsible person was chosen to represent the company. The representative was protected from shareholder interests in that he could not be dismissed without government approval and could implement plans without obtaining approval in the general shareholder meetings. Moreover, as a result of the "Designation System of Financial Institutions for Munitions Companies," firms were relieved of the pressure to obtain financing under this system, since financing was guaranteed by one or two financial institutions assigned to them by the Ministry of Finance.[11] In return for their privileged status, the munitions firms were obligated to conform to state plans, accommodate government requests concerning production, and assume responsibility for meeting production goals.

The state sought to reduce bureaucratic controls at the firm level by granting the firms greater leeway in achieving production targets. Restrictions and obligations imposed by previous laws were cancelled and special regulations and permits were granted. It also recognized profits as a legitimate means to ensure production. The government retained its low price policy, but was willing to subsidize losses incurred by firms and guarantee assistance and minimum profits when necessary.[12] The government essentially adopted the policy of supporting the most powerful and successful munitions firms. Each firm had a designated bank that ensured smooth and prompt funding. Moreover, the government helped develop a subcontracting system for each firm through the "Outline of Corporate Adjustment for Bolstering War-fighting Capacity" of June 1943. In addition, cooperative associations of subcontractors were formed to facilitate information flow and improve production methods.[13] The new strategy aimed to help the leading companies increase productivity by providing financial guarantees and backing within the overall framework of state production targets and goals.

The new policy toward industry represented a pragmatic adjustment, not a rejection of the New Order. The New Order for Industry was but one component of a broader vision that included New Orders for finance, labor, and science-technology. In this vision, the era of laissez-faire capitalism had been superceded by a new type of organized capitalism that recognized profits but within the confines of the national economy.

---

11. Okazaki, "The Japanese Firm under the Wartime Planned Economy," 371.
12. *Gunjushō*, "*Gunju kaisha hō kaisetsu*," in Hara Akira and Yamazaki Shirō, eds., *Gunju kankei shiryō* (Tokyo: Azuma shuppankai, 1997), 4–8.
13. Okazaki, "The Japanese Firm under the Wartime Planned Economy," 299.

The Bureaucratic New Order

As the Cabinet Planning Board report made clear, business was only part of the problem of the new control associations; the other problem was the bureaucracy. After stepping down from his post as vice minister of commerce, Kishi devoted his efforts to launching a "bureaucratic new order" (*kankai shintaisei*). From early 1941 Kishi and other reformists at commerce began to lay the groundwork for administrative reforms. Their efforts culminated in the reorganization of the economic bureaucracy and establishment of the Ministry of Munitions in 1943.

Reform bureaucrats had called for the streamlining of the bureaucracy and strengthening of the cabinet since the mid 1930s. At the Cabinet Research Bureau, Suzuki Teiichi drafted bold plans for the reorganization of the bureaucracy along the lines of the Manchukuo bureaucracy. From 1941, however, they no longer justified these reforms under the ideological banner of fascism or totalism, but in terms of the dictates of war and empire. Chastened by the recent attacks on the industrial new order, Kishi downplayed any relation between the new plans and previous efforts to reform the bureaucracy. He argued instead that the current measures were necessary to cut costs and increase efficiency for the war effort and dispatch civil servants to new administrative posts in Southeast Asia.[14]

The new bureaucratic order aimed at a complete overhaul of the bureaucratic system as it existed since Meiji. It targeted four areas: bureaucratic ethos, civil servant employment system, organizational structure, and duties and responsibilities. One of the main problems was that the bureaucratic structure failed to meet the demands of the advanced national defense state. According to Minobe Yōji, the Meiji bureaucracy had been conceived within the liberal framework based on the concept of a balance of power in which the various ministries competed against each other and checked each other's authority. Policy differences, turf battles, and other conflicts among the ministries were resolved by the various ministers and prime minister at the regular cabinet meetings. As a result of the increasing complexity of the state and the need for comprehensive, top-down planning and coordination, particularly in the economic realm, the traditional structure was no longer appropriate. Not only had the scope of administration expanded, but the very nature of administration was rapidly changing.[15]

Traditionally, Japanese bureaucrats derived their authority from their official status and tended toward a rule-based, passive, supervisory approach toward individual firms. Within each firm, labor, funds, and materials were organically managed and bureaucratic regulation of these individual components via its

---

14. Kishi and Noyori, "Senjika no keizai seisaku taidan, 52.
15. Minobe Yōji, "Keizai gyōsei kikō no kakushin," *Kaizō* (1941), 56.

various administrative organizations was kept to a minimum. Under the controlled economy, bureaucrats retained their old status-oriented and rule-based approach, while attempting to adopt a more active role in regulating and managing business. As a result, they disrupted the organic management of the firm. According to Minobe, the problem presented itself at various levels. For example, in order for the state to control the overall supply of funds, labor, and materials, it required firms to apply for licenses from different bureaucratic organizations in order to obtain the necessary inputs for production. The licensing system introduced new administrative burdens and obstacles in the production process, which in turn lowered productivity. In addition, numerous problems arose in coordinating the various economic functions and duties among the ministries. Although the commerce and agricultural ministries were in charge of production and distribution, they were dependent on the finance and welfare ministries for funds and labor. Moreover, among the economic ministries there was considerable overlap in areas of jurisdiction. For instance, the agricultural ministry was in charge of raw materials such as fats and oils or silk thread, while the commerce ministry handled the manufacture of these materials into soap and other fat-based products or silk goods.

Problems also arose when the bureaucracy tried to regulate the production and consumption aspects of the market. As the ministries became involved in managing the relationship between producers and consumers, they in turn adopted their respective viewpoints. In the case of the coal industry, the battles between the Ministry of Agriculture and Forestry and the Ministry of Commerce and Industry over pricing were fierce. The former, as the representative of coal producers, fought for a higher price; the latter tended to identify with the needs of distributors and consumers and argued for a lower price.[16] In other words, planning displaced the conflicts of interest and information flows from the market into the bureaucracy itself.

Kishi sought a fundamental reorientation of the bureaucracy away from its traditional, status-bound and rule-based approach toward a more task-oriented approach that focused on increasing productivity and performance. Hinting at the bold reforms to come, Kishi called for a general streamlining and strengthening of the bureaucracy and reduction of administrative personnel. He noted that despite numerous calls for bureaucratic streamlining under the previous cabinets, not a single department or bureau had been cut. If anything, in response to the need for wartime controls, the number of department sections and bureaus within the ministry had increased. Kishi warned that strengthening the

---

16. Kishi and Noyori, "Taidan kankai shintaisei ron," *Jitsugyō no sekai* 38, no. 9 (September 1941), 71–72.

bureaucracy did not mean an expansion of the number of bureaucratic offices along the old traditional lines. As in the creation of the economic new order, he explained, what was needed was a qualitative reform of the bureaucracy under the principle of prioritization (*jūtenshugi*) in order to channel precious manpower where it was most needed.[17]

The real force of Kishi's criticism of the bureaucracy was directed at the bureaucrats themselves. Reflecting on the problem of bureaucratic efficiency, he suggested that the structural issues represented only the tip, whereas the attitude of bureaucrats represented the root or source.[18] Namely, the main problem was the power mentality of bureaucrats or bureaucratic sectionalism. He noted that it would be impossible to establish a bureaucratic new order and raise efficiency unless the turf battles among bureaucrats were eliminated. As he put it, in the same way that business leaders must sacrifice their pursuit of private profits and work for the public interest, so bureaucrats must renounce their pursuit of jurisdictional power and focus on ways to improve the overall efficiency of the government.[19] As "servants of the emperor" (*heika no kanshi*), bureaucrats should place the public's welfare over their own impulse toward power. They were neither a privileged class nor enjoy superior status and should carry out their duties with a sincere and humble attitude.

Kishi admitted that bureaucrats not only needed to improve their relationship toward the public but also with each other as well. He reminisced about the early days when the bureaucracy was like a big family and bureaucrats exhibited a sense of personal responsibility and concern toward each other. Colleagues expressed concern over the frequent tardiness of a junior member, a colleague's overindulgence in alcohol, or his extramarital affairs. In recent years with the rapid expansion of duties, the bureaucracy had become a cold and impersonal place where department and section heads knew and cared little about the welfare of their staff and ministers. Moreover, with the frequent cabinet changes, new ministerial appointments were made every three to six months with the result that the incoming ministers were poorly informed about the work of their ministries and unable to skillfully manage their staff.[20] Kishi argued that what was needed were leaders who not only possessed intelligence and knowledge but also a warm character and personal touch. As part of an effort to improve the working environment, he called for higher compensation for bureaucrats,

---

17. Ibid., 73.

18. Kishi and Noyori, "Senjika no keizai seisaku taidan," 53.

19. Kishi Nobusuke, Akinaga Tsukizō, Kogane Yoshiteru, Nakano Yūrei, and Nakanishi Torao, "Sangyō shintaisei no shinro" *Kagakushugi kōgyō* (April 1941), 120.

20. Kishi and Noyori, "Senjika no keizai seisaku taidan," 58.

particularly at the middle and junior level. These salary increases could be paid for with funds made available from overall staff reductions in the bureaucracy.

Kishi's most radical proposal was to open up the civil servant employment system to the private sector in order to attract new talent and expertise. As he explained, the Meiji bureaucratic appointment ordinance had outlived its purpose of providing a regularized and impartial system of recruitment and training and cultivating an esprit de corps among civil servants. With the increase in scope and complexity of administration, particularly in the economic area, officials with technical and practical experience were urgently needed. Kishi believed that bureaucrats ought to be recruited not on the basis of passing the rigorous civil servants exam but on the basis of skill, knowledge, and practical experience. By abolishing this ordinance and eliminating the examination requirement, people from the private sector could become eligible for public office. Ultimately, Kishi believed that such reforms would encourage greater interchange and dialogue between the private and public sector and eliminate the status-oriented, rule-bound, power mentality of bureaucrats. In this regard, he saw the control associations as playing a pivotal role in encouraging a closer working relationship between government and business.[21]

Under the new Tōjō cabinet in November 1941, administrative reform became a top priority. As a tentative measure leaders established the Emergency Production Expansion Commission in November 1942. The commission was headed by the prime minister and comprised of the head of the Cabinet Planning Board and all key bureaucratic section and department chiefs connected with production expansion.[22] In order to strengthen policymaking at the executive level and cut through bureaucratic sectionalism and red tape, the Wartime Special Administration Law (*Senji gyōsei tokurei hō*) and Wartime Special Administration Powers Ordinance (*Senji gyōsei shokken tokurei*) were passed in March 1943. The former provided for the issuance of imperial ordinances for the purpose of expanding productive power that would overrule existing legislation prohibiting or controlling certain activities and permit intervention in areas under ministerial jurisdiction. The latter greatly increased the authority of the prime minister over the ministries with regard to the production of the five priority industries of iron and steel, coal, light metals, ships, and aircraft. The prime minister could order ministers to implement policies related to these materials regardless of existing laws or ministerial prerogative. He could also command a ministry to address problems outside of their jurisdiction or even to order the

---

21. Ibid., 76–77.

22. Cohen, *Japan's Economy in War and Reconstruction,* 70; Tsūshō sangyōsho, *Sangyō tōsei,* vol. 11, *Shōkō seisakushi* (Tokyo: Shōkō seisakushi kankōkai, 1964), 69.

transfer of bureaucrats to another ministry.[23] Furthermore, a Cabinet Advisory Council was established at the same time comprised of technocrats and seven leading industrialists including Mitsui executive Fujiwara Ginjirō and Riken president Ōkōchi Masatoshi. The council provided greater exchange and collaboration between bureaucrats and the private sector—something that reform bureaucrats had been advocating since the late 1930s.

Under the pressures of war, Kishi made significant advances toward realizing his vision of bureaucratic reform. In November 1943, he abolished the Cabinet Planning Board and consolidated the ministries of agriculture, commerce, communications, and railway into three new ministries: the Ministry of Munitions, the Ministry of Agriculture and Commerce, and the Ministry of Transport and Communications. The Munitions Ministry served as both a comprehensive planning organ and ministry for heavy industry. Its two most powerful bureaus were the General Mobilization Bureau and the Air Ordnance Bureau. The former absorbed the General Affairs Bureaus of the Cabinet Planning Board and Commerce Ministry and took over the planning functions of the Cabinet Planning Board. The latter oversaw all aspects of aircraft production. The army's and navy's aviation control associations were combined into the Aircraft Industries Control Association and came under control of the Munitions Ministry. The new ministry absorbed those bureaus of the commerce ministry related to heavy industry as well as the Electric Power Bureau of the Communications Ministry. Tōjō appointed himself minister of the new ministry, Kishi vice minister and effective head, and Shiina as General Mobilization Bureau chief.

Redefining "Rich Country, Strong Army"

As the war against China dragged on with no end in sight, the extent of Japan's overall national strength and readiness for war were called into question by pessimists at home and critics abroad. Akinaga acknowledged the formidable challenges facing Japan. As he explained in October 1941, from the standpoint of national income, Japan bore the triple economic burden of prosecuting the war in China, expanding productive power, and investing on the Asian continent. Already the China War had incurred sums exceeding 20 billion yen, more than ten times the cost of previous military conflicts such as the Russo-Japanese War (1.7 billion yen) and Manchurian invasion (2 billion yen). For 1941, the extraordinary military budget was 4.8 billion yen in addition to the General Account budget of 7.9 billion yen, of which more than 4 billion yen was earmarked for

---

23. Cohen, *Japan's Economy in War and Reconstruction,* 70; Tsūshō sangyō sho, *Sangyō tōsei,* 521–522.

production expansion-related costs and more than 1 billion yen for investment in Asia.[24] Based on such financial assessments, many observers, including American and British leaders, concluded that Japan would be brought to its knees.

Technocratic leaders, in contrast, believed that Japan had a good chance of prevailing against the larger, resource-rich nations allied against it. Their optimistic view was based on their new conception of national strength. As Akinaga and other military planners had argued, in the modern total wars, the definition of national power had changed. Economic power was but one component of national power. Two other factors—human labor and spiritual power—were equally important and without them, materials and funds would have no value. Japanese leaders boasted that their country was blessed with abundant labor power both in terms of its population growth rate and the excellence of the Yamato people, particularly with regard to its brain power. They were confident that the efficient organization and redeployment of labor to productive, war-related industries and steady population growth would overcome any labor shortages. Ultimately, they believed that more than labor power, the spiritual power or will power of the nation would determine Japan's fate. No matter how blessed a country was with material and human resources, they argued, a state could not display its total power without the spiritual mobilization of every member of the nation.[25]

Leaders claimed that the recent German military victories in Europe and its spirited offensive against England provided the basis for a positive reassessment of Japan's war potential. The new zaibatsu leader Nakano Yūrei reflected on the reasons for England's poor performance. From the standpoint of "rich country, strong army" England was a financially rich nation with access to abundant resources. A resource-poor country such as Germany should have been no match for England. But in the recent blitzkrieg, clearly other factors such as technical organization and mobilization strategy were decisive. The new type of war was based on a new type of thinking centered on materials and technology, not finance and diplomacy.[26] As Kishi explained, the meaning of "rich country, strong army" had changed. National wealth and power were no longer measured by a country's national income, but by the quantity and quality or precision of its materials and the ways in which they were organized and mobilized for national defense.[27] The challenge was to increase production through superior organization and eliminate the contradictions and inconsistencies in the production process.

24. Akinaga, "Idai naru kokuryoku no saikentō," 20.
25. Ibid. For a similar argument see Godō Takuo, ed., *Kokubō shigen ron* (Tokyo: Nihon hyōron, 1938), 8–9.
26. Kishi, Akinaga, et al., "Sangyō shintaisei no shinro," 102.
27. Ibid., 104–105.

Both Kishi and Nakano believed that in the new world order, economies would undergo a fundamental shift from a money-based economy to a materials-based economy. Such a shift reflected the dictates of the planned economy under which material balances and quotas, not prices and profits, served as the benchmark for economic activity. But more important, it highlighted the pivotal role of technology in the production process and in the creation of synthetic resources.

In order to bolster their claims about Japan's national strength, reform bureaucrats suggested a new theoretical approach toward measuring national wealth. In a roundtable of prominent Japanese economists in October 1941, Mōri argued that classical economic theory had become outdated in terms of both its assumptions and methodology. Until recently the nation's resources had been assessed by national income (in present day terms the total amount of goods and services produced in an economy), which was based on the individual's pursuit of self-interest. Mōri proposed to define national wealth (*kokumin shiryoku*) as the "total capital mobilization of the state" or "total productive power of state capital." State financial resources were distributed for public finance, consumption, and industry for the purpose of contributing toward the war economy and maximizing the efficiency of state planning. National wealth should not be assessed in the monetary terms of national income, which also included elements that did not directly contribute to the war economy, but rather in terms of their relative value or contribution toward fulfilling state plans. Reform bureaucrats incorporated this new view of national wealth into their "New Outline of Banking and Public Finance" (*Kinyū zaisei shin yōkō*).

Reform bureaucrats offered their own reformist methods of public finance accounting in order to calculate national wealth. Rather than dividing the budget into a General and Special Accounts, they proposed to create four categories within the General Accounts budget for official finance, reproduction or reinvestment, reserves and stockpiling, and welfare. Whereas the state would continue the traditional cost-benefit management approach toward regular day-to-day official finance, it would adopt what they referred to as the "long-term investment approach" toward the other three categories. In other words, they argued that welfare, production, and stockpile-related programs should be assessed in terms of their future long-term benefits and returns and not by their present worth. By rejecting the theoretical basis and methodology of foreign assessments of Japan's national wealth, reformists sought to show how the outside world underestimated the extent of Japan's true wealth and war preparation.

Finally, technocrats argued that national power should be understood in terms of its dynamic force. They believed that the synergistic energy derived from its material, human, and spiritual resources and the self-propelling momentum of Japan's advanced national defense state would determine victory in war. According to the government engineer Matsumae Shigeyoshi, the power of the national defense state should not be expressed in the static terms of the size of its air force or number of troops but rather in the dynamic terms of the state's ability to focus the energies of every aspect of society toward the goal of the perpetual expansion of productive power, technological advance, and increased efficiency both in terms of time, materials, and labor. Matsumae explained the dynamic nature of national power by likening the national defense state to a magnet whose force continually pulls the iron particles in one direction and in turn magnetizes them.[28] In another mechanical analogy, he compared the national defense state to a top:

> A top spins on its axis. The faster the top spins the more it stabilizes. When it spins at a very high speed, it attains a degree of stability by which motion and inertia become indistinguishable. As the rotational power gradually weakens, it begins to totter. At the end, when its rotational speed finally reaches zero, the top falls on its side. The so-called national defense state is a state with tremendous rotational force. Needless to say the essential idea behind the defense state is the dynamic rotation which concentrates the total power of the state, or the totality of the economy, the military, politics, and culture, at the center.[29]

The attempt to redefine national power in terms of such mechanical analogies and other intangible forms of spiritual and organizational power, potential national wealth, and "revisionist" accounting in the face of real material shortages, financial crisis, and human suffering revealed the moral compromises of Japan's technocratic leaders. The utter absurdity of Matsumae's analogy offered three insights into wartime technocratic leadership. First, it conveyed the deep disdain and contempt of Japan's wartime leaders for public opinion and discourse about politics and matters of life and death such as war. Second, the retreat into abstract formulations about spinning tops and magnets suggested a difference in degree, not essence, of the shallowness of the theoretical reasoning and rational formulations of technocrats. The above analogy offered a poignant caricature of the seemingly sophisticated, cosmopolitan theories about geopolitics, the new world order, and the national essence. More important, it revealed the alarming irresponsibility of Japan's wartime leaders and their inability or refusal to grapple with real issues determining their nation's fate.

The propaganda efforts of the reform bureaucrats did little to raise the level of official discourse on the war. At the forefront of the state's propaganda was

---

28. Matsumae Shigeyoshi, "Kōdō kokubō kokka," *Kōgyō kumiai* (October 1941), 15.
29. Ibid., 13–14.

Okumura, who assumed the post of deputy chief of propaganda at the Cabinet Information Bureau. Established in 1940, the new bureau replaced the Cabinet Information Committee and incorporated the Foreign Ministry's Information Department, the army's Newspaper Unit, the navy's Military Propaganda Department, the Home Ministry's Censorship Section, and the Communication Ministry's Wireless Section. The bureau possessed vast powers as reflected in its five sections created to handle planning and research; newspaper, publications, and broadcasts; foreign propaganda; censorship; and cultural propaganda. The president, Tani Masayuki, concurrently held the post of foreign minister from 1942. As deputy chief, Okumura personally shaped public opinion through his control over the mass media and foreign and domestic propaganda. He was also a key figure behind the state's control over scholarship and research and cultural and literary activities.[30]

In his radio broadcasts, Okumura portrayed the "Greater East Asia War" as fundamentally an ideological war. According to Okumura, the concept of an ideological war had first emerged during World War I. The United States, England, and France launched an aggressive thought war against Germany. As a result of the liberal powers' effective use of all types of propaganda medium, including newspaper, radio, film, music, public speech, and the arts, they were not only able to win the propaganda war, but also the ideological war, and established modern democracy and internationalism as the accepted basis of the world order. Japan sought to free itself of the ideological hegemony of the liberal powers through the Manchurian Incident. Despite Anglo-American pressure to force Japan to conform to the liberal order via its harsh criticism of the Manchurian invasion and its Washington arms treaties, Japan defied the liberal status quo and joined up with Germany and Italy to promote a new world order.[31] In his speeches, Okumura proclaimed that the ultimate goal of the war was "the annihilation of the former ideology of England and the United States which dominated the old order of Greater East Asia and the construction of a new order of 'eight corners under one roof' based on the founding spirit of the country of our emperor Japan."[32]

Okumura doubted that the liberal powers had the spiritual will and unity to sustain a long battle against the spiritually united Axis powers. He viewed America's liberal system, in which state interests took a back seat to private interests, as particularly ill-suited for war. Okumura and Mōri, who joined the Cabinet Information Bureau in January of 1942, questioned whether the American materialist lifestyle could adjust to the needs of a national defense state.

30. Hata Ikuhiko, *Kanryō no kenkyū* (Tokyo: Kōdansha, 1983), 139.
31. Okumura Kiwao, *Sonnō jōi no kessen* (Tokyo: Ōbunsha, 1943), 345–356.
32. Ibid., 345.

Due to Japan's occupation of Southeast Asia, the United States was no longer able to obtain sufficient supplies of resources such as sugar, oil, gasoline, tin, and rubber. They confidently predicted that Americans would not be able to endure the material shortages and that its "blockaded economy" would give rise to economic chaos.[33]

Reform bureaucrats fought the ideological battle not only over the air waves but in society at large. They warned that the enemies of Japan were not only the Americans and British, but all those who harbored Western liberal ideas or promoted them in some form. In order to fight the enemies "within," Okumura and his colleagues at the Cabinet Information Bureau, together with the police, launched an aggressive campaign to purge the remnants of liberal thought in Japan. In December 1942, the bureau launched the Great Japan Patriotic Press Association (*Dai Nihon genron hōkokukai*). In his opening address at the general meeting to establish the new association, Okumura stated that the ideological war went beyond the "strategic, propagandistic, and technical aspects" and aimed at changing the people's beliefs. He hoped to not only eradicate the enemies' liberal ideas but also to eventually correct their worldview. However, he emphasized, the urgent task at hand was the elimination of the American and English spirit among the Japanese people and the revival of the "Yamato spirit."[34]

The Greater East Asia Co-Prosperity Sphere

The concept of the Greater East Asia Co-Prosperity Sphere served as a complex ideological matrix that brought together various strands of Japanese techno-cratic and right-wing thinking. It fused managerial concepts of the multilateral business structure, führer principle, and *Grossraumwirtschaft* (greater Japanese economic sphere or *kōiki keizai*) with geopolitical ideas of an "organic state" that required living space and Japanese pan-Asianist visions of Asian liberation into a fascist vision of empire. These strands of thought mutually reinforced each other in their common vision of a hierarchical, organic, functionalist community. It was a product of the collaboration of the military, pan-Asianists, and ultrana-tionalists, as well as technically minded professionals, including economic and regional planners, geographers, and engineers.[35]

---

33. Mōri Hideoto, "Dai Tōa sensō to Ei-Bei sensō keizairyoku," *Kokusaku hōsō* (March 1942), 8–14; Okumura Kiwao, "Haisen Beikoku no kokumin seikatsu," *Shūhō* (June 1942), 22–30.

34. A copy of his speech is contained in Okumura, *Sonnō jōi no kessen,* 389–395.

35. Peter Duus, "The Greater East Asian Co-Prosperity Sphere: Dream and Reality," *Journal of Northeast Asian History* 5 no. 1 (June 2008), 134–154; Eri Hotta, *Pan-Asianism and Japan's War* (New York: Palgrave Macmillan, 2007).

The concept can be seen as the product of decade-long attempts of technocrats to promote their occupational status and interests beginning with the establishment of Manchukuo. As a social group, technocrats sought legitimacy within Japanese society but could not obtain it by appealing to technical expertise alone. They formed alliances with other social groups, on whose behalf they offered their technical skills. Technocrats in turn embraced the ideology and interests of groups they represented and helped produce a larger political rationale that encompassed the various interests. In order for Japanese technocrats to justify their leadership role in Asia's New Order, they needed to generate a new vision of empire that could appeal to various groups in Japan and other Asian countries.

Technocrats made clear the material rewards to the military and business that Asia provided. Already in the fall of 1939 Hoshino, then chief of the General Affairs Agency of Manchukuo, had enthusiastically spoken about a new "Eastern Autarky," whose construction rested on the twin goals of the development of heavy industry and the achievement of self-sufficiency in foodstuffs.[36] As a result of the China War, he wrote, Japan had acquired control over China's resources and could now anticipate the achievement of near complete autarky. Materials needed for heavy industry included its "four cornerstones" coal, iron, oil, and sulphur, as well as limestone, copper, lead, aluminum, rubber, salt, and wood. The Japan-Manchuria-China bloc could supply all essential resources except for oil, copper, and rubber.

Hoshino confidently predicted that any remaining deficiencies could be corrected through the creation of synthetic substitutes. Oil shortages could be alleviated through the rapid increase in production of Manchurian oil shale, liquefaction of coal, and excavation of recently discovered oil fields in Western Manchuria. Synthetic rubber could be produced from limestone, coal, and electric power, of which Manchuria and China had an abundant supply. As for copper, copper ore was recently discovered in Manchuria; in addition, aluminum could be substituted for copper. Self-sufficiency in agricultural produce, especially foodstuffs, was to be achieved through the reclamation of Manchuria's abundant land, particularly in the northern part of the country. Finally the human labor needed to develop these resources was plentiful. Hoshino also noted that the utilization of the various peoples of the East—70 million Japanese, 30 million Koreans, 30 million Manchurians, and several hundred million Chinese provided a reserve army for the defense of the bloc.[37]

---

36. Hoshino Naoki, "Shin tōyō arutaruki no kensetsu," *Kaizō* (October 1939); Hoshino Naoki, Kawai Yoshinari, Matsuda Reisuke, Kōda Noboru, Matsushima Kagami, and Yamada Katsuto, "Manshūkoku keiei no mokuhyō o kataru zadankai," *Jitsugyō no Nihon* (January 1940), 24–33.

37. Hoshino, "Shin Tōyō arutaruki no kensetsu," 23–28.

Following its attack on Pearl Harbor, Japan came into possession of the precious materials that the Japan-Manchuria-China bloc lacked. In a broadcast to the nation on December 19, 1941, Kishi reported on the vast resources of Asia. The Philippines possessed superior iron ore and abundant flax, as well as coal, chrome and manganese ore. Malaysia was the world's largest producer of rubber, tin, iron ore, coal, manganese, tungsten, fluorite, and bauxite. The Dutch East Indies had rich supplies of oil, rubber, tin, coal, iron ore, bauxite, copper, manganese, lead, zinc, chrome, tungsten, mercury, bismuth, and anchimon. As for the South Seas, Kishi described them as a treasure house of minerals that had yet to be mined. He noted that there were only a few resources in which Greater East Asia was not sufficient. Through science and technology, Japan would create substitutes for these resources.[38]

In early 1942, following the string of Japanese victories over the Allied Powers, Vice Commerce Minister Shiina acknowledged that some people likened Japan's recent acquisition of the vast resources of Southeast Asia to a "cat being given a whale."[39] While admitting that such views of Japanese policy were probably inescapable, Shiina and his colleagues sought to portray the war not as an imperialist one, in which Japan would feast upon the vast resources of Asia, but as a moral and constructive war for the benefit of Asia. Appealing to Asian liberation and brotherhood, they argued that the current war was a "holy war" (*seisensō*) fought by Japan to replace the "egotistical," "power-oriented blocs" of the Western colonial leaders with a Japan-centered "moral bloc" that promoted Asian prosperity and culture.

At the same time, the current battle was depicted as a "war of construction" (*kensetsu sensō*) in which Japan was building a *Grossraumwirtschaft* reflecting the modern trend toward national land planning and great power blocs. Technocrats argued, from the standpoint of economic rationality, that the weak, backward countries of Asia could not thrive independently outside of a larger regional bloc. Only through the synergies and economies of scale of such a bloc, along with the technological leadership of Japan, could Asia hope to compete with the West. Moreover, by describing the war as a "hundred year war" technocrats emphasized Japan's long-term commitment to the Asian region. In the new era of multiyear planning, they explained, the first phase of construction would focus on obtaining essential raw materials needed for military victory against the Allies. This phase would eventually be followed by the long-term development of basic, civilian industries in Asia.

---

38. Kishi Nobusuke, "Kōdō kokubō keizai taisei no shinten to dai tōa no shigen," December 12, 1941 broadcast, in *Nihon senji keizai no susumu michi* (Tokyo: Kenshinsha, 1942), 3–5.

39. Shiina, "Nanpō shinshutu no kamae," 20, 23.

The Greater East Asian Co-Prosperity idea put forth an alternative, ideological basis and a new unifying, organizational principle to articulate the multiple military, political, economic, cultural, and ethnic ties between Japan and Asia. As a "pan idea," it was based on the geopolitical theory that the world would be divided into pan-regions consisting of four large economic spheres centered on the "core" industrial regions of the United States, Germany, the Soviet Union, and Japan. Within the bloc, "co-prosperity" would replace the Wilsonian ideal of "open door" in East Asia. In place of the liberal principles of "self-determination" and "self-interest" of the individual Asian countries within the international economy, reformists advanced the principle of "coexistence" of the Asian peoples within a self-sufficient bloc. Its organizational basis would not be free trade based on a country's comparative advantage in natural resources or profitable market strategy, but rather the organic, hierarchical, functionalist principles of totalism and the multilateral business organization in which each member country, according to its ability (*kaku minzoku no bun ni ōjite*), contributes its raw materials, labor, capital, or technological expertise for the benefit of the bloc as a whole.

Technocrats emphasized that Japan would not replace the West as the new imperialist power in Asia. Rather, Western capitalist colonial "exploitation" of Asia would give way to mutual "co-prosperity" of a region whose liberation would lead to its increased wealth and power. Ultimately, though, they justified Japanese leadership of the Asian sphere to themselves not in terms of superior Japanese technology, but in terms of the Japanese geopolitical notion of "Great Japan" (*dai Nihon*), in which Japan was a selected, superior organism that was entitled to grow at the expense of other Asian countries.

Forging a New National Identity

Technocrats claimed that through the establishment of the Greater East Asia Co-Prosperity Sphere, Japan was transforming itself from a small, resource-poor, maritime trading nation to a resource-rich, self-sufficient continental power. They began to promote a new vision of "continental Japan" (*tairiku Nihon*) following the seizure of Manchuria. The change in national identity reflected a new approach toward empire. Technocrats approached the Manchurian venture and the Greater East Asia Co-Prosperity Sphere as an expansive, task-oriented, technical project that promoted a productivist vision of mutual gain, especially for business and labor.

Technocrats conceived of Japan's empire in the spatial-strategic concepts of Japan-Manchuria bloc, Japan-Manchuria-China bloc *Grossraumwirtschaft*, living sphere, and co-prosperity sphere. They saw Japan's position shifting from a

peripheral nation in the capitalist world order to a core nation within the concentrically arranged regional bloc. Planners described the co-prosperity sphere
as consisting of a "core sphere" composed of Japan, Manchuria, North China,
the lower Yangtze region, and the Soviet-occupied north coastal region, a "lesser
sphere" composed of the core sphere and Eastern Siberia, China, Indochina, and
the South Pacific, and a "greater sphere," which included the lesser sphere as well
as Australia, India, and the Pacific Islands.[40] The latter represented no more than
the outer boundary or peripheral sphere of the Japan-Manchuria-China bloc. As
with capital in the domestic New Order, the physical possession of Asia's terri-
tory was not a necessary condition for its control. What was needed was superior
organization and the reorientation of the people toward the common goals of
Asian co-prosperity.

As in the Manchurian venture, reformists sought to devise ethnic nationalist
principles of rule to win the hearts and minds of the occupied people and avoid
rule by force. Whereas in the case of Manchuria, Japanese leaders brandished the
slogans of kingly way and ethnic harmony, now they promoted the principle of
"eight corners under one roof" (hakkō ichiu). The new slogan defined Japan's
central position within the new Asian empire based on the principles of Japan's
hierarchical family system composed of the ruling house or clan and subordinate
branch houses.[41] It advanced the notion of a moral hierarchical order in Asia led
by Japan and its emperor.

With regard to the co-prosperity sphere, technocrats walked a fine line between presenting Japan as the defender of Asia against Western imperialism and
justifying Japanese hegemony in the region. In Manchuria they appealed to geopolitical theory to support the military's continental policy, arguing that it represented a lifeline that would enable Japan to grow and become a continental
power. Now in justifying the new Asian bloc, reformists promoted the geopolitical concept of "living sphere" to explain the military's dual strategy of northern
and southern advance.

In his formulation of the East Asian cooperative body in the late 1930s, Mōri
had distinguished between Japan's reformist continental policy in north China
and its liberal, imperialist maritime policy in central and south China. Now he
modified his position and argued that the two living spaces of the Asian continent and the Pacific Ocean were being united into a "homogenous single space."
For example, in 1940 Mōri Hideoto argued that the Pacific Ocean had taken
on a new significance and was becoming the foundation of a new world order;
he suggested that "with regard to the historical stage of the life struggle of the

---

40. Kawahara, *Shōwa seiji shisō kenkyū,* 303–305.
41. Kikakuin, *Dai tōa kensetsu no kihon yōkō* (Tokyo: Dōmei tsūshinsha, 1943), 15.

Japanese ethnic people, [we] have finally discovered the possibility of organizing the waters of the Pacific Ocean, together with our land, into a living sphere."[42]

In a public broadcast a month after Japan's declaration of war against the Allied powers, Mōri proclaimed that Japan's possession of both a continental and maritime base placed it in the most optimal position geopolitically to win the war. England was a maritime power but lacked a continental base, whereas the continental powers Germany and the United States both lacked maritime bases. He predicted that Germany, although it was on its way to establishing a European continental state via the European war, would be handicapped geopolitically because of its lack of a maritime base. Japan, in contrast, with its recent acquisition of the vast resources of Asia, would be able to build an undefeatable "greater East Asian economy of co-prosperity."[43]

National Land Planning

The new sphere represented a turn toward "national land planning" (*kokudo keikaku*) whose basic goals and principles were laid out in the Cabinet Planning Board's "Outline for the Establishment of National Land Planning" (*Kokudo keikaku settei yōkō*). It also provided a great planning opportunity for technocrats. National land planning was conceived as part of Konoe's New Order movement, but went beyond other New Order plans in incorporating a broader and more comprehensive spatial dimension to planning. National planning took place at the new Total War Research Institute. Founded in September 1940, it was designed as a research and training institute to promote research on total war primarily through collaborative research projects with government and business.[44] First headed by Cabinet Planning Board president Hoshino Naoki, its staff included Akinaga Tsukizō, Mōri Hideoto, Okumura Kiwao, Minobe Yōji, Sakomizu Hisatsune, Yamazoe Risaku from the Ministry of Agriculture, and Ōshima Hiroaki from the Home Ministry.

Japanese technocrats described national planning as the most advanced form of state planning. According to one technocrat, national planning went beyond the narrowly conceived "production technology" (*seisan gijutsu*) of the Soviet, Manchurian, German, and Japanese five- and four-year plans. While these plans merely sought to meet limited, short-term targets for increasing production in

---

42. Kamakura Ichirō, "Nichi-Doku-I dōmei to kongo no Nihon: Taiheiyō kūkan no seikaku ka-kumei," *Chūō kōron* (November 1940), 35.

43. Mōri, "Dai Tōa sensō to Ei-Bei sensō keizairyoku," 14.

44. Kawahara, *Shōwa seiji shisō kenkyū*, 303–307; Furukawa, *Shōwa senchūki no sōgō kokusaku kikan*, 194–195.

industry and agriculture by temporary measures such as extending labor time or installing new equipment within a given geographical setting, national planning represented a new type of "construction technology" (*kensetsu gijutsu*) in which officials took a long-term—one-hundred-year—approach and sought the optimal geographical location of industries within the bloc.[45] Now, the state sought to determine the most efficient distribution of the various facilities of the economy, population, culture, and society in order to promote the comprehensive development, use, and preservation of the native land in accordance with the state's goal.[46]

National planning in Japan, as in other industrialized countries, sought to address the problems arising from liberal capitalism: overconcentration of industries in metropolitan areas, reckless destruction of valuable farm land and deforestation for the sake of industrial growth, and inefficient location of industries. It also implied a social critique of the "metropolitan character" of modern capitalism and the social evils associated with it.[47] Moreover, it was justified from the standpoint of national defense as a means to maximize industrial production, reduce vulnerability of industries to foreign air attack, and ensure self-sufficiency in foodstuffs.

All advanced industrialized states faced similar challenges of urbanization, metropolitan congestion, and decline of the countryside as a result of the industrial revolution in the nineteenth century. National land planning was first introduced and promoted by British planners as part of the movement for regional and urban planning. It was advocated as a means to decrease overpopulation and congestion in the major metropolitan areas by promoting satellite cities and towns, incorporating green belt areas, building a nationwide transportation network system, and formulating plans for regional growth. In contrast to the liberal type of national land planning focusing on suburban development, the authoritarian regimes of Soviet Russia, Germany, Italy, and Japan looked to national land planning primarily as a way to expand national productive power. The Soviet five year plans, German four year plans, Manchurian five year plans, and Japanese four year plans represented the first steps toward authoritarian national land planning.

Japanese planners classified national land planning in the various industrialized countries according to two general criteria. First, the state adopted

---

45. Yokota Shūhei, *Kokudo keikaku to gijutsu* (Tokyo: Shōkō gyōseisha, 1944), 41–44.

46. See the "Outline for the Establishment of National Land Planning" (*Kokudo keikaku settei yōkō*) of September 24, 1940, in Kikakuin kenkyūkai, *Kokubō kokka no kōryō*.

47. Hatano Sumio, "'Tōa shinchitsujo' to chiseigaku," in Miwa Kimitada, ed. *Nihon no 1930 nendai* (Tokyo: Sōryūsha, 1980), 24–27.

either authoritarian planning from above or democratic planning from below depending on whether it had a liberal or totalist political system. Second, depending on the particular developmental circumstances and history of a country, the state pursued the goal of either redesigning existing areas (*kokudo saihenseishugi*) or developing new land (*kokudo shinkōshugi*). Among countries that possessed undeveloped frontier land, the United States pursued grassroots planning from below, reflecting its liberal tradition, whereas Soviet Russia imposed centralized planning from above in accordance with its authoritarian political system.[48] Among those smaller countries that lacked open uncultivated land and were constrained to focus on restructuring developed areas, England attempted a bottom-up type liberal planning to address the social problems of industrialization, whereas Germany pursued top-down planning primarily for the purposes of national defense. Ultimately though, Japanese planners viewed the liberal system as an obstacle to true national land planning. They argued that since liberal countries did not tolerate top-down planning, they could only partly implement national land planning from below. Planning of the vast undeveloped resources in the United States stopped at the regional level because the state was not strong enough to restrain freedom and coordinate the various interests at the local and regional level. In terms of restructuring metropolitan areas as in England, the challenges were multiplied. Suburban planning in England never took off because of the state's inability to tackle the source of urban congestion: the laissez-faire economy, which permitted uncontrolled economic and urban development devoid of an overall planning authority and vision.[49]

Reformists pointed out that national planning was not individual planning expanded to the national level but rather the task of "determining the order of the land and striving toward its comprehensive functioning at the highest efficiency level."[50] For this reason, they believed that totalist regimes were best suited to carry out national land planning. Moreover, among totalist states, they believed that the Japanese case was unique because Japan possessed both the challenges of reorganization of its native land and frontier development of its East Asian empire. The Japanese state's goals were to build a national defense system in Japan that incorporated strategic spatial planning for defense, establish an autarkic sphere in East Asia to secure resources for Japan, address Japan's social

---

48. Ishikawa Hideaki, *Nihon kokudo keikaku ron* (Tokyo: Hachigensha, 1941), see chapter 1, esp. his schematic presentations on 14 and 32.

49. Ibid., 16, 7.

50. Ibid., 11.

problems of urbanization resulting from rapid industrialization, and coordinate the various plans in a comprehensive way.

Japanese scholars have highlighted the nativist rhetoric and idealism in Japan's vision of the Greater East Asia Co-Prosperity Sphere as an important factor in leading to its decision to attack the United States. They accuse Japanese wartime leaders of self-deception, abandonment of realism for idealism leading to a "warped national consciousness," and irresponsible and ineffective political leadership.[51] However, I would argue that their idealism was expressed in not only the nativist rhetoric and visions of Asian co-prosperity bandied about by the right, but in a techno-fascist vision that combined technological advance with pan-Asianism, geopolitics, and progressive notions of a socially inclusive, functionalist society. Japanese technocrats were convinced of the bankruptcy of the liberal order and believed that Japan stood at the forefront of a new world order in which fascism as a third way was replacing liberalism and communism. They promoted a highly technological idealism that expressed confidence in Japan's technological prowess and its self-appointed mission in Asia. At the same time, this vision reflected the pragmatic, occupational interests of technocrats who coveted a leadership role in Japan and its empire

When one considers the enormous investment of energy, time, and resources into plans for Manchukuo, Japan's New Order, and its expanding Asian empire, it is difficult to dismiss the concept of the Greater East Asia Co-Prosperity Sphere as simply an empty slogan or rhetorical cover. Certainly the contrast between the high-sounding goal of "co-prosperity" and the reality of war and brutal subjection of Asian peoples to Japanese interests was acute. One cannot ignore the underlying ethnic chauvinist assumption of the Greater East Asia Co-Prosperity Sphere: that Japan, whether on geopolitical or nativist grounds, was destined to expand its power and promote its interests at the expense of its Asian neighbors. But it is less useful to reduce the concept to simply a problem of Japanese chauvinism or duplicity. The Greater East Asia Co-Prosperity Sphere expressed not only the dreams of pan-Asianists and ultranationalists, but also the technocratic concepts and visions born out of wartime efforts to theorize about Japan's modern state and its relationship to Asia that have important implications for the postwar era.

---

51. Maruyama, *Thought and Behaviour in Modern Japanese Politics,* 94; Miwa Kimitada, "Japanese Policies and Concepts for a Regional Order in Asia, 1938–1940," in James W. White, Michio Umegaki, and Thomas R.H. Havens, eds., *The Ambivalence of Nationalism: Modern Japan Between East and West* (Lanham, MD: University Press of America, 1990).

EPILOGUE

From Wartime Techno-Fascism
to Postwar Managerialism

This study suggests that the main political faultline in wartime Japan was not between militarists and peace-loving civilians, but between advocates of technocratic reform and defenders of the capitalist status quo. Membership in each camp cut across the traditional affiliations of the military, bureaucracy, business, labor, political parties, and academia. The driving force for reform was the professional class, which included military planners, reform bureaucrats, the new zaibatsu, progressive intellectuals, labor party leaders, and government engineers. On the opposing side were conservative elites such as the old zaibatsu, mainstream party politicians, traditional bureaucrats, and imperial advisors. Some leaders such as Konoe Fumimaro tried to straddle both sides, a position that resulted in deep personal anguish, confusion, and belated peace initiatives to the Americans in the early 1940s.

The vicious battles between reformists and conservatives throughout the 1930s and early 1940s provide important insights into prewar Japan. The vigorous defense of zaibatsu autonomy and elite privilege, especially in the face of the Economic New Order, suggests the tenacity of the liberal capitalist principles of profit, private property, and business autonomy among Japan's ruling class. Not until the military's attack on Pearl Harbor did business finally come around to support the government's control policies, and only after being guaranteed substantial subsidies and perks.

The struggles also reveal the radical vision and innovative strategies of Japanese technocrats. Although technocrats portrayed themselves as impartial, non-ideological servants of science-technology and the state, they were among the key supporters of techno-fascism and the march toward empire and war. Their solution to the problems of the 1930s was to achieve an ideal type of techno-fascism that would resolve the tensions and contradictions of modern society and secure a leadership role for technocrats. Techno-fascism promoted technology and progressive social reform, through ethnic chauvinism, violence, and political oppression. Its vision of a New Order in Japan and Asia was progressive and reactionary, defensive and aggressive, inclusive and exclusive, egalitarian and hierarchical, and rational and irrational. Its complex, synthetic approach represented a new type of politics of transcendence that reached out to many groups.

This synthetic approach helps explain the support of techno-fascism among the left-wing technical intelligentsia. Technocrats on the left and the right shared a common desire to overthrow the old ruling class and create a new type of society. Technology represented the meeting point for both sides. In addition to advancing the right's pan-Asianist agenda, it promised to increase efficiency and productivity, improve the conditions and wages of the working class, and elevate Japanese technocrats to their coveted place in society. The inability of postwar progressive intellectuals and socialist politicians to acknowledge the left's complicity and attraction to fascism remains an important and insufficiently explored topic in the debates on Japanese war responsibility and war memory.

The origins of Japanese techno-fascism can be located in both the interwar crisis of capitalism and the new opportunities presented by modern technology. Its political and conceptual strategies were first formulated in the introduction of state planning, beginning with state-guided industrial rationalization and the founding of Manchukuo. Conceptually, techno-fascism represented an authoritarian form of rationalization that sought to achieve freedom through control, innovation through organization, autonomy via community, and status via hierarchy. This eclectic approach was used in the construction of Manchukuo, where Kwantung Army officers and elite Japanese bureaucrats collaborated with diverse groups and interests and forged a new antiliberal vision of state and society under the pan-Asianist slogans of ethnic harmony, kingly way, and co-prosperity. From the late 1930s reform bureaucrats drew on these experiences in Manchuria and the model of Nazi Germany to formulate their own techno-fascist vision and political strategy for Japan.

Techno-fascism, in its ideal political and economic form, could not be realized during the war. Technocrats failed to overcome the power of the zaibatsu, Home Ministry, throne, and military factions. The biggest obstacles were the Meiji Constitution and the restrictive peace preservation and public order laws, which enshrined the principle of private property and restricted popular political mobilization. Only through a charismatic leader and mass party or dictatorship

could conservative resistance, bureaucratic sectionalism, and military factionalism be overcome. Technocrats lacked a charismatic leader and the emotional
appeal and material incentives to mobilize the people. In the end, only a total
war and the prospect of near self-annihilation fostered national unity, but not
on their own terms.

The transition from wartime techno-fascism to postwar managerialism was
characterized by, first and foremost, a fundamental shift in political and economic goals. Following defeat and occupation, Japan renounced war and empire
and reentered the international community as a capitalist trading partner com-
mitted to peace and democracy. The state-centered plan of the advanced national
defense state was replaced by a society-centered plan aimed at the creation of a
middle-class consumer society. The motors of growth for the postwar state were
no longer the military, munitions industries, and empire, but rather the middle
class, civilian industries, and international trade. Japan's economic recovery and
growth was facilitated by Cold War tensions and the outbreak of the Korean War
in 1950, which brought about a "reverse course" in U.S. policy toward strengthening Japan and the sudden rise of overseas demand for Japanese goods.

In the first decade of the postwar period, however, the old battle lines between
reformist technocrats and status quo conservatives could still be detected. Following the initial period of conservative rule under Yoshida Shigeru, a prewar
diplomat and member of the old guard, punctuated by the brief socialist and
coalition cabinets of Katayama Tetsu and Ashida Hitoshi, former wartime technocrats quickly consolidated their forces. Technocratic advance was marked by
the successive stages of political rehabilitation of their leader, Kishi Nobusuke:
first by his release from Sugamo Prison in 1948, then his formation of the Liberal Democratic Party (LDP) in 1955, and finally, his assumption to the prime
ministership in 1957.

The American occupation greatly contributed to the postwar ascendance of
civilian technocrats. SCAP removed the wartime obstacles to technocratic rule
by dismantling the Home Ministry, breaking up the zaibatsu, stripping the emperor of power, and replacing the Meiji constitution with one that allowed popular political participation. Moreover, following the "reverse course" from the late
1940s, SCAP drew heavily on the expertise of former wartime technocrats to
foster economic growth.

From the perspective of global historical trends, the ascendance of Japanese
technocrats seems overdetermined. Japan, like Germany, reentered the modern
world during the heyday of technocracy, when mature industrial societies were
reaching an advanced stage of organizational integration. The new "postindus-
trial" or "managerial" societies of the 1950s and 1960s were described in terms
of the "techno-structures" of capitalism, the rise of "organization man," and the

"end of ideology."[1] The old symbols of technocratic modernity—New Dealism, Italian Fascism, Nazism, and Stalinism—gave way to the new symbols of
the American "military-industrial complex," Gaullism, *Wirtschaftswunder,* and
de-Stalinization. Social scientists such as Herbert Marcuse and Alain Touraine
warned that postwar societies were introducing a new form of totalitarian domination whose political and ideological orientations were elusive and difficult to
grasp using the traditional concepts of class and political power.[2] Like other industrial nations, Japan responded to and was shaped by the common challenges
of the postwar period: reconstruction (of one's own or a defeated country),
decolonization, the new age of the mass consumer and its related technologies
(transistor radio, television, and household appliances, as well as advertising and
marketing), and the Cold War.

From this perspective, the years immediately preceding and following Japan's
surrender in August 1945 appear as a temporary break in an otherwise continuous transwar history of Japanese technocratic planning that began after World
War I. This history, however, does not view the prewar bureaucracy as a monolithic force, nor is it premised on the long tradition of "bureaucratism" that pre-
ceded Japan's emergence as a modern state. Rather it focuses on the fundamental
ways in which the Japanese state was transformed from within by a small but
influential group of bureaucrats.

The central figure in this transformation was Kishi Nobusuke. Under Kishi,
the reform bureaucrats spearheaded the movement to establish a new institutional framework for technological advance and economic growth and promote
a new generation of planners and technically minded bureaucrats. Their legacy
was profound as evidenced by the numerous institutional and personnel links
between prewar and postwar planning.[3] The wartime planning apparatus centered on the Ministry of Commerce, Cabinet Planning Board, Ministry of Munitions, and control associations was carried over into the postwar period in the
form of the Ministry of International Trade and Industry (MITI), Economic Stabilization Board (ESB), Economic Planning Agency (EPA), and Keidanren. These
organizations were staffed by former wartime technocrats such as Shiina Etsu-
saburo, two-time MITI minister; Sakomizu Hisatsune, EPA director-general; and
Wada Hiroo, EPA director and Minister of Agriculture.

---

1. Galbraith, *The New Industrial State;* William H. Whyte, *The Organization Man* (Garden City,
NY: Doubleday, 1957); Daniel Bell, *The End of Ideology: On the Exhaustion of Political Ideas in the
Fifties* (Glencoe, IL: Free Press, 1960).

2. Herbert Marcuse, *One-Dimensional Man* (Boston: Beacon Press, 1964), 3; Alain Touraine, *The
Post-Industrial Society* (London: Wildwood House, 1969), 71.

3. For a study of prewar and postwar continuities of Japanese industrial policy, see Johnson,
MITI and the Japanese Miracle.

Kishi's personal imprint can be seen in basic aspects of Japan's postwar managerial state. His efforts throughout the 1930s and early 1940s to blur the lines between private and public and foster close ties between business and the state came to fruition in such postwar institutions as the various deliberation councils, the Keidanren, and the Liberal Democratic Party. More than any other institution, the LDP cemented the links between the bureaucracy and business. As LDP secretary-general and then prime minister from 1957 to 1960, Kishi built the "1955 System" of the LDP's one-party rule through his extensive ties in business, politics, and the bureaucracy. Dominated by the mainstream faction of retired bureaucrats, the LDP has served as the key link between Japanese bureaucrats and the public. It has provided them with political legitimacy in the eyes of the people as well as practical, real-world experience.

In addition, Kishi transformed Japan's conservative leadership by shifting power away from the prewar conservative elites represented by Yoshida to middle-class professionals. As scholars have noted, Kishi incorporated shadier aspects into conservative politics, including financial dealings with underworld business leaders and right-wing figures such as Sasakawa Ryōichi and Kodama Yoshio.[4] Kishi also advanced the interests of the broader group of middle-class professionals, including technical bureaucrats, former new zaibatsu, small- and medium-sized businesses, and socialists. He maintained ties to former Social Masses Party leaders such as Asō Hisashi, Nishio Suehiro, and Miwa Jusō, a close personal friend who legally represented him during his imprisonment. Following the defeat of his new political party, the Japan Reconstruction Federation in 1952, Kishi considered joining the Socialist Party.[5]

As prime minister, Kishi cultivated close relations with the United States and renewed the highly controversial U.S.–Japan Security Treaty in 1960. This strategy reflected neither a repentant stance toward Japanese wartime aggression nor an embrace of American-style democracy. It was the product of cool calculation of how to use the escalating Cold War and Japan's ally to its best advantage. Cold War politics conveniently deflected attention away from the problem of war responsibility and Japanese aggression in Asia. Kishi never wavered from his prewar technological worldview and vision of Japan standing shoulder to shoulder with the technological superpowers the United States, the Soviet Union, and Germany. Nor did he express regret over Japan's occupation of Manchuria and plans to create a Greater East Asia Co-Prosperity Sphere. In his forward to the postwar collection of reminiscences of Manchukuo planners entitled *Ah, Manchuria,* Kishi emphasized the youthful idealism and aspirations of Manchukuo

---

4. Samuels, *Machiavelli's Children,* 225–249.

5. Kishi, Yatsugi, and Itō, eds., *Kishi Nobusuke no kaisō* (Tokyo: Bungei shunjū, 1981), 99.

as the "hope of Asia" and Japan's important contribution in developing Manchuria's economy. He and his wartime colleagues continued to view Manchukuo as a technocrat's dream and a sincere attempt to liberate and develop Asia.[6] However, their program to liberate Asia was, in reality, a plan to secure Japan's ruling position in the future world order by using Asia as a pawn and ultimately a victim in the game of great power politics.

---

6. Manshū kaikoshū kankōkai, *Aa Manshū,* see foreword by Kishi Nobusuke.

Bibliography

Aikawa Haruki. "Shintaisei to gijutsu no soshikika." *Gijutsu hyōron* (January 1941), 13–18.

Aikawa Shūji. "Gijutsu no kaihō to rōdō no rinen." Minobe Yōji Documents, L 1:80.

Akinaga Tsukizō. "Idai naru kokuryoku no saikentō." *Jitsugyō no sekai* (October 1941), 19–21.

———. "Senjika ni okeru seisan zōshin." *Nihon nōritsu* 2, no. 4 (April 1943), 14–17.

Akinaga Tsukizō, Hoshino Naoki, Mōri Hideoto, Shiina Etsusaburō et al. "Manshūkoku keizai no genchi zadankai." *Tōyō keizai shinpō* (October 1936), 26–39.

Andō Yoshio. "Nihon senji keizai to shin kanryō." In *Shimin shakai no keizai kōzō*, edited by Takahashi Kōhachirō, Andō Yoshio, and Kondō Akira, 456–478. Tokyo: Yūhikaku, 1972.

Arendt, Hannah. *The Origins of Totalitarianism*. New York: Harcourt Brace, 1976.

Arisawa Hiromi. *Sangyō dōin keikaku*. Tokyo: Kaizōsha, 1934.

———, ed. *Shōwa keizaishi*. Tokyo: Nihon keizai shinbunsha, 1976.

Asada Kyōji and Kobayashi Hideo, eds. *Nihon teikokushugi no Manshū shihai: Jūgonen sensōki o chūshin ni*. Tokyo: Jichōsha, 1986.

Ayukawa Yoshisuke. "Kaiketsu Ayukawa Yoshisuke." *Nihon hyōron* (July 1936), 421–438.

———. "'Nissan' Manshū ichū e no hōfu—Watakushi no yarō to suru Manshū sangyō kaihatsu shin soshiki." *Tōyō keizai shinpō* (November 1937), 27–32.

———. "Nissan no Manshū shinshutsu." *Keizai zasshi daiyamondo* (November 1937), 114–115.

———. "Atarashiki sekai no tenkai to sekai keizai oyobi sangyō." *Kyōiku kenkyū* (January 1938), 24–32.

———. "Yo no tōsei keizai kan." *Jitsugyō no Nihon* (August 1938), 18–22.

———. "Hito to jigyō." *Keizai ōrai* (May 1939), 280–285.

———. "Watakushi no rirekisho." Vol. 24. Tokyo: Nihon keizai shinbunsha, 1970.

———. "Nissan kontsuerunu no seiritsu." Interview. *Shōwashi e no shōgen*. Vol. 2, edited by Andō Yoshio. Tokyo: Hara shobō, 1993, 106–128.

Ayukawa Yoshisuke sensei tsuisōroku hensan kankōkai, ed. *Ayukawa Yoshisuke sensei tsuisōroku*. Tokyo: Ayukawa Yoshisuke sensei henshū kankōkai, 1968.

Baba Akira. *Nitchū kankei to gaisei kikō*. Tokyo: Hara shobō, 1983.

Baba Tsunego, Rōyama Masamichi, Katayama Tetsu et al. "Kanryō no chōryo to fuhai o kataru zadankai." *Nihon hyōron* 11, no. 11 (November 1936), 84–118.

Bain, Foster. "Manchuria: A Key Area." *Foreign Affairs* 25, no. 1 (October 1946), 106–117.

Barkai, Avraham. *Nazi Economics: Ideology, Theory, and Policy*. Oxford: Berg, 1990.

Barnhart, Michael A. *Japan Prepares for Total War: The Search for Economic Security, 1919–1941*. Ithaca: Cornell University Press, 1987.

Barshay, Andrew E. *The State and Intellectual in Imperial Japan*. Berkeley: University of California Press, 1988.

Ben-Ghiat, Ruth. "Italian Fascism and the Aesthetics of the 'Third Way.'" *Journal of Contemporary History* 31 (1996), 293–316.

——. *Fascist Modernities.* Berkeley: University of California Press, 2001.

Bender, Ursula. *Technik: Technischer Fortschritt und soziolökonomische Zusammenhänge bei Friedrich von Gottl-Ottlilienfeld.* Frankfurt: Peter Lang Verlag, 1985.

Berezin, Mabel. *Making the Fascist Self.* Ithaca: Cornell University Press, 1977.

Berger, Gordon M. *Parties out of Power, 1931–1941.* Princeton: Princeton University Press, 1977.

——. "The Three Dimensional Empire: Japanese Attitudes and the New Order in Asia, 1937–1945." *The Japan Interpreter* 12, nos. 3–4 (1979), 355–383.

Berle, Adolf A., and Gardiner C. Means, *The Modern Corporation and Private Property.* New York: Macmillan, 1933.

Berman, Marshall. *All That Is Solid, Melts in the Air: The Experience of Modernity.* New York: Penguin Books, 1982.

Betz, Horst. "How Does the German Historical School Fit In?" *History of Political Economy* 20 no. 3 (fall 1998), 409–430.

Bisson, Thomas A. "Aspects of Wartime Economic Control in Japan." Secretariat Paper No. 2. New York: Institute of Pacific Relations, 1945.

Bōeichō bōei kenshūjo. *Rikugun gunju dōin.* Vol. 1. *Keikakuhen.* Tokyo: Asagumo shinbunsha, 1967.

Borg, Dorothy, and Shumpei Okamoto, eds. *Pearl Harbor as History.* New York: Columbia University Press, 1973.

Brady, Robert A. *The Rationalization Movement in German Industry: A Study in the Evolution of Economic Planning.* Berkeley: University of California Press, 1933.

Brooks, Barbara. *Japan's Imperial Diplomacy: Consuls, Treaty Ports, and War in China, 1895–1938.* Honolulu: University of Hawaii Press, 2000.

Burleigh, Michael, and Wolfgang Wipperman. *The Racial State: Germany 1933–1945.* New York: Cambridge University Press, 1991.

Buck-Morss, Susan. "Envisioning Capital: Political Economy on Display." *Critical Inquiry* 21 (winter 1995), 434–467.

Burnham, James. *The Managerial Revolution: What Is Happening in the World.* Bloomington, IN: Greenwood Press, 1941.

Butow, Robert C. *Tojo and the Coming of the War.* Stanford: Stanford University Press, 1961.

Calder, Kent E. *Crisis and Compensation.* Princeton: Princeton University Press, 1988.

Chiwitt, Ulrich. *Wirtschaft und Leben: Eine philosphische Analyse der Wirtschaftslehre Friedrich von Gottl-Ottlilienfeld.* Essen: Die Blaue Eule, 2000.

Cohen, Jerome B. *Japan's Economy in War and Reconstruction.* Minneapolis: University of Minnesota Press, 1949.

Coleman, Samuel K. "Riken from 1945 to 1948: The Reorganization of Japan's Physical and Chemical Research Institute under the American Occupation," *Technology and Culture* 31, no. 2 (April 1990), 228–250.

Coox, Alvin D. *Nomonhan: Japan against Russia, 1939.* 2 vols. Stanford: Stanford University Press, 1985.

Crowley, James B. "Japanese Army Factionalism in the Early 1930s." *Journal of Asian Studies* 21, no. 3 (May 1962), 309–326.

——. *Japan's Quest for Autonomy: National Security and Foreign Policy, 1930–1938.* Princeton: Princeton University Press, 1966.

——. "Intellectuals as Visionaries of the New Asian Order." In *Dilemmas of Growth in Prewar Japan,* edited by James W. Morley, 319–373. Princeton: Princeton University Press, 1971.

Djilas, Milovan. *The New Class.* San Diego: Harcourt Brace Jovanovich, 1985.

Doak, Kevin M. *Dreams of Difference: The Japan Romantic School and the Crisis of Modernity.* Berkeley: University of California Press, 1994.

——. "Ethnic Nationalism and Romanticism in Early Twentieth-Century Japan." *Journal of Japanese Studies* 22, no. 1 (winter 1996), 77–103.

——. *A History of Nationalism in Modern Japan.* Leiden: Brill, 2007.

Dower, John W. *War without Mercy: Race and Power in the Pacific.* New York: Pantheon Books, 1986.

——. "The Useful War." Reprinted in John W. Dower, *Japan in War and Peace: Selected Essays, 19–32.* New York: New Press, 1993.

Drea, Edward J. *The 1942 Japanese General Election: Political Mobilization in Wartime Japan.* International Studies, East Asian Series Research Publication, no. 11. Lawrence, KS: University of Kansas, 1979.

Drucker, Peter F. *The End of Economic Man.* New York: John Day, 1939.

——. *The New Society.* New York: Harper & Brothers, 1950.

Duara, Pransenjit. *Sovereignty and Authenticity: Manchukuo and the East Asian Modern.* Oxford: Rowman & Littlefield, 2003.

Duus, Peter. "The Reaction of Japanese Big Business to a State-Controlled Economy in the 1930s." *Rivista Internazionale di Scienze Economiche e Commerciali* 31, no. 9 (September 1984), 819–831.

——. *The Abacus and the Sword: The Japanese Penetration of Korea, 1895–1910.* Berkeley: University of California Press, 1998.

——. "The Greater East Asian Co-Prosperity Sphere: Dream and Reality." *Journal of Northeast Asian History* 5, no. 1 (June 2008), 143–154.

Duus, Peter, Ramon H. Meyers, and Mark Peattie, eds. *Japan's Wartime Empire.* Princeton: Princeton University Press, 1996.

Duus, Peter, and Daniel Okimoto. "Fascism and the History of Pre-War Japan: The Failure of a Concept," *Journal of Asian Studies* 39, no. 1 (November 1979), 65–76.

Earle, Edward Mead, ed., *Makers of Modern Strategy: Military Thought from Machiavelli to Hitler.* New York: Atheneum, 1966.

Eley, Geoff. "What Produces Fascism: Pre-Industrial Traditions or a Crisis of the Capitalist State?" In Geoff Eley, *From Unification to Nazism: Reinterpreting the German Past, 254–282.* Boston: Allen and Unwin, 1986.

Ellul, Jacques. *The Technological Society.* New York: Vintage Books, 1964.

Ezawa Shōji. *Nihon kokudo keikaku no kisō riron.* Tokyo: Nihon hyōronsha, 1942.

Fletcher, William Miles. *The Search for a New Order.* Chapel Hill: University of North Carolina Press, 1982.

Friedmann, Georges. "Technological Change and Human Relations." *The British Journal of Sociology* 3, no. 2 (June 1952), 95–116.

Fritzsche, Peter. "Nazi Modern." *Modernity/Modernism* 3, no. 1 (January 1996), 1–21.

Fujiwara Yutaka. *Manshūkoku tōsei keizai ron.* Tokyo: Nihon hyōronsha, 1942.

Fukui Kōji. *Sei toshite no keizai.* Tokyo: Kōbundō shoten, 1937.

Furukawa Takahisa. "Kakushin kanryō no shisō to kōdō." *Shigaku zasshi* 99, no. 4 (April 1990), 1–38.

——. *Shōwa senchūki no sōgō kokusaku kikan.* Tokyo: Furukawa kōbunkan, 1992.

Furumi Tadayuki. *Wasureenu Manshūkoku.* Tokyo: Keizai ōraisha, 1978.

Galbraith, John K. *The New Industrial State.* Boston: Houghton Mifflin, 1967.

Gao, Bai. "Arisawa Hiromi and His Theory for a Managed Economy." *Journal of Japanese Studies* 20, no. 1 (winter 1994), 115–153.

——. *Economic Ideology and Japanese Developmentalism.* New York: Cambridge University Press, 1997.

Garon, Sheldon. *The State and Labor in Modern Japan.* Berkeley: University of California Press, 1987.

——. *Molding Japanese Minds: The State in Everyday Life.* Princeton: Princeton University Press, 1997.

Gentile, Emilio. "Impending Modernity: Fascism and the Ambivalent Image of the United States." *Journal of Contemporary History* 28, no. 1 (January 1993), 7–29.

Gerschenkron, Alexander. *Economic Backwardness in Historical Perspective.* Cambridge: Harvard University Press, 1962.

Gluck, Carol. *Japan's Modern Myths.* Princeton: Princeton University Press, 1985.

Godō Takuo. "Dai tōa sensō to nōritsu zōshin"—Nihon nōritsu taikai ni okeru kōen." *Nihon nōritsu* 2, no. 4 (April 1943), 2–10.

——, ed. *Kokubō shigen ron.* Tokyo: Nihon hyōron, 1938.

Götz Aly, and Susanne Heim. *Architects of Annihilation.* London: Weidenfeld & Nicholson, 2002.

Gordon, Andrew. *Labor and Imperial Democracy in Prewar Japan.* Berkeley: University of California Press, 1991.

Gottl-Ottlilienfeld, Friedrich von. *Vom Sinn der Rationalisierung.* Jena: Gustav Fischer, 1929.

Gouldner, Alvin W. "Metaphysical Pathos and the Theory of Bureaucracy." *American Political Science Review* 49, no. 2 (June 1955), 496–507.

——. *The Dialectic of Ideology and Technology: The Origins, Grammar, and Future of Ideology.* New York: Seabury Press, 1976.

Griffin, Roger. *The Nature of Fascism.* London: Routledge, 1991.

Gurvitch, Georges. *The Social Frameworks of Knowledge.* New York: Harper Torch, 1972.

Habermas, Jürgen. *Toward a Rational Society.* Boston: Beacon Press, 1971.

Hadley, Eleanor M. *Antitrust in Japan.* Princeton: Princeton University Press, 1970.

Hamaguchi Yūko. *Nihon tōchi to higashi-Ajia shakai.* Tokyo: Keisō shobō, 1996.

Hara Akira. "1930 nendai no Manshū keizai tōsei seisaku." In *Nihon teikokushugika no Manshū,* edited by Manshūshi kenkyūkai, 1–114. Tokyo: Ochanomizu shobō, 1972.

——. 'Manshū' ni okeru keizai tōsei seisaku no tenkai: Mantetsu kaisō to Mangyō setsuritsu o megutte. In *Nihon keizai seisakushi ron,* edited by Andō Yoshio, 209–296. Tokyo: Tokyo daigaku shuppankai, 1976.

——. "Senji tōsei keizai no kaishi." In *Nihon no rekishi.* Vol. 20. *Kindai,* 218–268. Tokyo: Iwanami kōza, 1981.

——. "Japan: Guns before Rice." In *The Economics of World War II,* edited by Mark Harrison, 224–267. Cambridge: Cambridge University Press, 1998.

Hara Akira and Yamazaki Shirō, eds. *Gunjushō kankei hōki kaisetsu.* Tokyo: Gendai shiryō shuppan, 1997.

Hara Yoshihisa. *Kishi Nobusuke: Kensei no seijika.* Tokyo: Iwanami shoten, 1995.

——. *Kishi Nobusuke shōgenroku.* Tokyo: Mainichi shinbunsha, 2003.

Harootunian, Harry. *Overcome by Modernity.* Princeton: Princeton University Press, 2000.

Hashikawa Bunzō. "Kakushin kanryō." In *Gendai Nihon shisō taikei.* Vol. 10. *Kenryoku no shisō,* edited by Kamishima Jirō, 251–273. Tokyo: Chikuma shobō, 1965.

——. "Shin kanryō no seiji shisō." In *Kindai Nihon seiji shisō no shosō,* edited by Hashikawa Bunzō, 296–304. Tokyo: Miraisha, 1968.

——. "Kokubō kokka no rinen." In *Kindai Nihon seiji shisōshi.* Vol. 2, edited by Hashikawa Bunzō and Matsumoto Sannosuke, 232–251. Tokyo: Yūhikaku, 1970.

——. "Tōa shinchitsujo no shinwa." In *Kindai Nihon seiji shisōshi.* Vol. 2, edited by Hashikawa Bunzō and Matsumoto Sannosuke, 352–367. Tokyo: Yūhikaku, 1970.

Hashimoto Jurō. "Kyōdai sangyō no kōryū." In *Nihon keizaishi.* Vol. 6. *Nijū kōzō,* edited by Nakamura Takafusa and Odaka Kōnosuke, 81–131. Tokyo: Iwanami shoten, 1989.

Hata Ikuhiko. *Kanryō no kenkyū*. Tokyo: Kōdansha, 1983.

Hatano Sumio. "'Tōa shinchitsujo' to chiseigaku." In *Nihon no 1930 nendai*, edited by Miwa Kimitada, 14–47. Tokyo: Sōryūsha, 1980.

Hatch, Walter, and Kozo Yamamura. *Asia in Japan's Embrace: Building a Regional Production Alliance*. Cambridge: Cambridge University Press, 1996.

Hauner, Milan. *What Is Asia to Us? Russia's Asian Heartland Yesterday and Today*. London: Routledge, 1992.

Haushofer, Karl. *Taiheiyō chiseigaku*, trans. Satō Sōichirō. Tokyo: Iwanami shoten, 1942.

Havens, Thomas R.H. *Farm and Nation in Modern Japan: Agrarian Nationalism, 1870–1940*. Princeton: Princeton University Press, 1974.

Hawley, Ellis W. *The New Deal and the Problem of Monopoly*. Princeton: Princeton University Press, 1966.

Heeger, Gerald A. "Bureaucracy, Political Parties, and Political Development." *World Politics* 25, no. 4 (July 1973), 600–607.

Hein, Laura. "Growth Versus Success: Japan's Economic Policy in Historical Perspective." In *Postwar Japan as History*, edited by Andrew Gordon, 99–122. Berkeley: University of California Press, 1993.

———. *Reasonable Words, Powerful Men*. Berkeley: University of California Press, 2004.

Herf, Jeffrey. *Reactionary Modernism: Technology, Culture, and Politics in Weimar and the Third Reich*. New York: University of Cambridge Press, 1986.

Hirano Ken'ichirō. "Manshūkoku kyōwakai no seijiteki tenkai: Fukusū minzoku kokka ni okeru seijiteki antei to kokka dōin." In *"Konoe shintaisei" no kenkyū*, edited by Nihon seiji gakkai, 231–283. Tokyo: Iwanami shoten, 1972.

———. "The Japanese in Manchuria, 1906–1931: A Study of the Historical Background of Manchukuo." Ph.D. diss., Harvard University, 1983.

Hiroshige Tetsu. *Kagaku no shakaishi—kindai Nihon no kagaku taisei*. Tokyo: Chūō kōronsha, 1973.

Homma Shigeki. "Senji keizai hō no kenkyū." *Shakai kagaku kenkyū* 25, no. 6 (March 1974), 1–56.

Hoshino Naoki. "Shin tōyō kensetsu no dōgiteki konkyo." *Kyōiku kenkyū* (February 1939), 3–7.

———. "Shin tōyō arutaruki no kensetsu." *Kaizō* (October 1939), 23–28.

———. *Mihatenu yume*. Tokyo: Daiyamondosha, 1963.

———. *Hoshino Naoki shi danwa sokkiroku*. Tokyo: Naiseishi kenkyūkai, 1964.

Hoshino Naoki, Kawai Yoshinari, Matsuda Reisuke et al. "Manshūkoku keiei no mokuhyō o kataru zadankai." *Jitsugyō no Nihon* (January, 1940), 24–33.

Hoshino Naoki and Kojima Seiichi. "Keizai yokusan tōji." *Kaizō* (November 1940), 203–225.

Hosokawa Ryūichirō. *Kishi Nobusuke*. Tokyo: Jiji tsūshinsha, 1986.

Hotta, Eri. *Pan-Asianism and Japan's War*. New York: Palgrave Macmillan, 2007.

Ienaga Saburō. *The Pacific War, 1931–1941*. New York: Pantheon, 1978.

Iguchi Haruo. *Unfinished Business: Ayukawa Yoshisuke and U.S.–Japan Relations, 1937–1985*. Cambridge, MA: Harvard University Asia Center, 2003.

Imai Kazuo. *Kanryō*. Tokyo: Yomiuri shinbunsha, 1953.

Imai Seiichi and Itō Takashi, eds. *Gendaishi shiryō*. Vol. 44, no. 2. *Kokka sōdōin: Seiji*. Tokyo: Misuzu shobō, 1974.

Imura Tetsuo, ed. *Kōain kankō tosho-zasshi mokuroku*. Vol. 17. *Jūgonen sensō jūyō bunken shirīzu*. Tokyo: Fuji shuppan, 1994.

———, ed. *Mantetsu chōsabu: Kankeisha no shōgen*. Tokyo: Ajia keizai kenkyūjo, 1996.

Inaba Hidezō. "Kanryō toshite no Wada Hiroo." *Kankai* 3 (February 1977), 176–186.

International Military Tribunal for the Far East. *Transcript of Proceedings: General Index of the Record of the Defense Case through the Tri-Partite Pact Section of the Pacific Phase, 16,998–24,758.* IPS Document no. 0008. Tokyo, 1947.

Ishibashi Tanzan. Ōkōchi Masatoshi, Matsui Haruo, et al. "Nihon no shigen o kataru kai." *Kagakushugi kōgyō* (June 1939), 164–182.

Ishidō Kiyotomo and Taniyama Toshitada, eds. *Tokyo teidai shinjinkaiin no kiroku.* Tokyo: Keizai jūraisha, 1976.

Ishikawa Hideaki. *Nihon kokudo keikaku ron.* Tokyo: Hachigensha, 1941.

Itagaki Yoichi. *Seijikeizaigaku no hōhō.* Tokyo: Nihon hyōronsha, 1942.

Itō Kinjirō. "Shin-kyū kanryō no shōtai." *Chūō kōron* Li. 9 (September 1936), 117–129.

Itō Takashi. "Shōwa jūsannen Konoe shintō mondai kenkyū oboegaki." In *"Konoe shintaisei" no kenkyū,* edited by Nihon seiji gakkai. Tokyo: Iwanami shoten, 1972.

———. "'Kyokko itchi' naikakuki no seikai saihensei mondai—Shōwa jūsannen Konoe shintō mondai kenkyū no tame ni," Part 1. *Shakai kagaku kenkyū* 24, no. 1 (August 1972), 56–130.

———. "'Kyokko itchi' naikakuki no seikai saihensei mondai," Part 2. *Shakai kagaku kenkyū* 25, no. 4 (February 1974), 59–147.

———. "The Role of Right-Wing Organizations in Japan." In *Pearl Harbor as History: Japanese-American Relations, 1931–1941,* edited by Dorothy Borg and Okamoto Shumpei, 487–509. New York: Columbia University Press, 1975.

———. *Shōwa jūnendaishi danshō.* Tokyo: Tōkyō daigaku shuppankai, 1981.

———. "Akinaga Tsukizō kenkyū oboegaki." Reprinted in Itō Takashi, ed., *Shōwaki no seiji* (zoku), 215–234. Tokyo: Yamakawa shuppankai, 1993.

———. "Mōri Hideoto ron oboegaki." Reprinted in Itō Takashi, *Shōwaki no seiji* (zoku), 235–260. Tokyo: Yamakawa shuppankai, 1993.

Itō Takeo. *Life along the Manchurian Railway: The Memoirs of Itō Takeo,* trans. Joshua A. Fogel. New York: M. E. Sharpe, 1988.

Johnson, Chalmers. *MITI and the Japanese Miracle: The Growth of Industrial Policy, 1925–1975.* Stanford: Stanford University Press, 1982.

———. *Japan: Who Governs? The Rise of the Developmental State.* New York: W. W. Norton, 1995.

Johnston, B. F. *Japanese Food Management in World War II.* Stanford: Stanford University Press, 1953.

Jones, F. C. *Manchuria since 1931.* New York: Oxford University Press, 1949.

Kamakura Ichirō (pen name of Mōri Hideoto). "'Tōa kyōseitai kensetsu no shojōken: Chōki kensetsu no mokuhyō." *Kaibō jidai* (October 1938), 23–30.

———. "'Tōa ittai' toshite no seiji ryoku." *Kaibō jidai* (November 1938), 6–11.

———. "Jiken dai-yonki wa seiji o tenkaisu." *Kaibō jidai* (December 1938), 70–77.

———. "Kokumin soshiki to tōa kyōdōtai no fukabunsei." *Kaibō jidai* (January 1939), 22–28.

———. "Chūgoku no 'kōsen kenkoku' o hihansu." *Kaibō jidai* (February 1939), 4–15.

———. "Tōa kyōdōtai to gijutsu no kakumei." *Kaibō jidai* (March 1939), 4–12.

———. "Nihon kokumin keizai no keisei to seiji." *Kaibō jidai* (April 1939), 25–32.

———. "Kokumin keizai to rieki." *Kaibō jidai* (May 1939), 83–89.

———. "Gijutsu no kaihō to seiji." *Kaibō jidai* (September 1939), 4–8.

———. "Tōsei keizai no hinkon no gen'in: Shizenryoku ka soshikiryoku ka." *Kaibō jidai* (December 1939), 16–21.

———. "Handō o kokufukusuru seiji." *Kaibō jidai* (January 1940), 4–12.

———. "Kokumin ishiki to seiji." *Kaibō jidai* (February 1940), 15–19.

——. "Jihen kansui no ishiki to taisei." *Kaibō jidai* (March 1940), 27–33.

——. "Chūshōteki na bukka to gutaiteki na bukka." *Kaibō jidai* (April 1940), 4–35.

——. "Senji keizai no shindankai." *Kaibō jidai* (May 1940), 31–35.

——. "Nihon tōa sekai no jidaiteki chitsujo." *Kaibō jidai* (July 1940), 12–18.

——. "Seiji-genri-seikatsu (seisaku, soshiki): Kokumin sōryoku taisei no keisei ni tsuite." *Kaibō jidai* (August 1940), 4–13.

——. "Nichi-Doku-I dōmei to kongo no Nihon: Taiheiyō kūkan no seikaku kakumei." *Chūō kōron* (November 1940), 34–42.

——. "Shina jihen to Ōshū sensō to no mitchaku." *Kaibō jidai* (November 1940), 31–35.

——. "Seiji ishiki to kagaku gijutsu suijun." *Gijutsu hyōron*, 18, no. 1 (January 1941), 24–26.

——. "Shinwa o motsu minzoku." *Kaibō jidai* (April 1941), 22–26.

——. "Hitotsu no kotae toshite no senji seikatsu sōdansho." In *Senji seiji keizai shiryō*, edited by Kokusaku kenkyūkai. Vol. 1, 8–9. Tokyo: Hara shobō, 1982.

Kamakura Ichirō and Miki Kiyoshi. "Ashita no kagaku Nihon no sōzō." *Kagakushugi kōgyō* (January 1941), 186–207.

Kamei Kan'ichirō. "Shina jihen no kisō kōzō." *Nihon hyōron* (December 1937), 70–86.

——. "Atarashiki sekaikan ni tatsu Nachisu Doitsu." *Bungei shunjū* (June 1938), 106–117.

——. "Doitsu no sangyō seishin." *Kagakushugi kōgyō* (August 1938).

——. "Kokumin no tō no hitsuyō." *Nihon hyōron* (May 1939), 186–192.

——. *Nachisu kokubō keizai ron*. Tokyo: Tōyō keizai shuppanbu, 1939.

——. "Kōa dantai tōchi no kōsō." *Chūō kōron* (March 1941), 92–101.

——. *Kamei Kan'ichirō danwa sokkiroku*. Tokyo: Nihon kindai shiryō kenkyūkai, 1969.

Kamei Kanichirō, Akamatsu Katsumaro, Nakano Tomio, et al. "Shintaisei no zenro o kataru," *Nihon hyōron* (December 1940).

Kan Tarō. Interview. In Nakamura Takafusa, Itō Takashi, and Hara Akira, eds., *Gendaishi o tsukuru hitobito*. Vol. 1, 225–262. Tokyo: Mainichi shinbunsha, 1971.

Kaneko Hiroshi. "Zentaishugi keizaigaku no nikeikō—Gottoru to Shupan." *Kokumin keizai zasshi* 65, no. 2 (August 1938), 35–48.

Kantōgun. *Manshū jihen jisshi*. Tokyo: Nittō shoin, 1932.

Kantōgun sanbōbu. *Kensetsu tojō no Manshūkoku*. Vol. 1. Kantōgun sanbōbu, 1933.

Kaplan, Alice Y. *Reproductions of Banality: Fascism Literature, and French Intellectual Life*. Minneapolis: University of Minnesota Press, 1986.

Kashiwara Heitarō, Minobe Yōji, Sakomizu Hisatsune, and Mōri Hideoto. "Zadankai: Kakushin kanryō—shintaisei o kataru." *Jitsugyō no Nihon* (January 1941), 52–67.

Kasza, Gregory J. *The State and the Mass Media in Japan, 1918–1945*. Berkeley: University of California Press, 1988.

——. *The Conscription Society: Administered Mass Organizations*. New Haven: Yale University Press, 1995.

——. "Fascism from Above? Japan's Kakushin Right in Comparative Perspective." In *Fascism outside Europe: The European Impulse against Domestic Conditions in the Diffusion of Global Fascism*, edited by Stein Ugelvik Larsen. Boulder, CO: Social Science Monographs, 2001, 183–232.

Katakura Tadashi. *Manshūkoku keizai seisaku no genjitsu to shōrai ni tsuite*. Tokyo: Nihon jitsugyō kyōkai, 1935.

Katakura Tadashi and Furumi Tadayuki. *Zasetsushita risō koku: Manshūkoku kōbō no shinsō*. Tokyo: Gendai bukkusu, 1967.

Katayama Tetsu. "Denryoku kokuei no mokuhyō to tsūshinsho an." *Nihon hyōron* (special edition: October 1936), 387–392.

Katō Shinkichi. "Manshū ni okeru minzoku kyōwa no mondai," in *Manshū mondai ni kansuru shiken.* Tokyo: Nittō shoin, 1932.

Katsuta Teiji. *Nihon zentaishugi keizai no seikaku.* Tokyo: Jitsugyō no Nihonsha, 1940.

Kawahara Hiroshi. *Ajia e no shisō.* Tokyo: Kawashima shoten, 1968.

———. *Shōwa seiji shisō kenkyū.* Tokyo: Waseda daigaku shuppansha, 1979.

Kawahara Hiroshi, Asanuma Kazunori, Takeyama Morio, Hamaguchi Haruhiko, Shibata Toshio, and Hoshino Akiyoshi, eds. *Nihon no fuashizumu.* Tokyo: Yūhikaku, 1979.

Kawanishi Seikan. *Tōa chiseigaku no kōsō.* Tokyo: Jitsugyō no Nihonsha, 1942.

Kerde, Ortrud. "The Ideological Background of the Japanese War Economy: Visions of the 'Reformist Bureaucrats.'" In *Japan's War Economy,* edited by Erich Pauer, 23–38. London: Routledge, 1999.

Kido Kōichi. *Kido Kōichi nikki.* 2 vols. Tokyo: Tokyo daigaku shuppankai, 1966.

Kikakuin kenkyūkai. *Kokubō kokka no kōryō.* Tokyo: Shinkigensha, 1941.

———. *Dai tōa kensetsu no kihon yōkō.* Tokyo: Dōmei tsūshinsha, 1943.

———. *Tōseikai no honshitsu to kinō.* Tokyo: Dōmei tsūshinsha, 1943.

Kimijima Kazuhiko. "Kōkōgyō shihai no tenkai." In *Nihon teikokushugi no Manshū shihai: Jūgonen sensōki o chūshin ni,* edited by Asada Kyōji and Kobayashi Hideo, 547–674. Tokyo: Jichōsha, 1986.

Kinmonth, Earl. "The Mouse that Roared: Saito Takao, Conservative Critic of Japan's 'Holy War' in China." *Journal of Japanese Studies* 25, no. 2 (summer 1999), 331–360.

Kishi Nobusuke. "Ōshū ni okeru sangyō gōrika no jissai ni tsuite." *Sangyō gōrika* (January 1932), 27–67.

———. "Jūyō sangyō tōsei hō kaisetsu." *Kōgyō keizai kenkyū.* Vol. 1 (April 1932), 51–76.

———. "Sangyō gōrika undō ni arawaretaru keiken kōkan." *Kōgyō keizai kenkyū* (July 1932), 99–120.

———. "Chūshō kōgyō keinyū gaikyō." *Kōgyō keizai kenkyū.* Vol. 3 (January 1933), 143–156.

———. "Sangyō gōrika yori tōsei keizai e." *Sangyō gōrika.* Vol. 12 (April 1934), 1–40.

———. "Manshū kenkoku rokushūnen ni tsuite." *Shina jihō* (March 1938), 96–99.

———. "Shin tōa kensetsu e no kakugo." *Kōgyō kumiai* (January 1940), 5–10.

———. "Manshū keizai no genjō ni tsuite." *Tōa keizai kōenroku: Tōa keizai kenkyūjo kaisetsu kinen.* Tokyo shōka daigaku, June 24, 1940.

———. "Shōgyō shinchitsujo to shinshōgyō rinri." *Jitsugyō no Nihon* (October 1940), 36–37.

———. "Atarashiki shōnindō." *Yūben* (November 1940), 24–29.

———. "Watakushi wa naniyue jikan o yametaka." *Jitsugyō no sekai* (February 1941), 48–49.

———. *Man-Shi sangyō no jitsujo.* Tokyo: Nihon kōgyō kurabu, 1941.

———. "Senjika no nōritsu zōshin." *Nihon nōritsu* 1, no. 1 (June 1942), 15–21.

———. "Shōgyō saihensei to shōgyōsha no shimei." *Kokusaku hōsō* (November 1942), 11–13.

———. *Nihon senji keizai no susumu michi.* Tokyo: Kenshinsha, 1942.

———. "Shisei de ugokanu mono wa nai: Watakushi no jinsei o kettei zuketa Shōin sensei no kotoba." Speech at Bunyūkai on May 9, 1980.

———. *Kishi Nobusuke kaikoroku: Hoshu gōdō to Anpo kaitei.* Tokyo: Kōsaidō shuppan, 1983.

———. *Waga seishun: Oitachi no ki, omoide no ki.* Tokyo: Kōsaidō shuppan, 1983.

Kishi Nobusuke, Akinaga Tsukizō, Kogane Yoshiteru, Nakano Yūrei, and Nakanishi Torao. "Sangyō shintaisei no shinro." *Kagakushugi kōgyō* (April 1941), 100–123.

Kishi Nobusuke and Imai Hisao. "Kenka Kishi Nobusuke: Kanryō ichidai." *Kankai* (November 1976), 104–112.

Kishi Nobusuke and Kamei Kan'ichirō. "'Tōa no shinro o ronzu' taidankai." *Bungei shunjū* (March 1939), 174–183.

Kishi Nobusuke and Noyori Hideichi. "Kishi Nobusuke to Noyori Hideichi to no jikyoku taidan." *Jitsugyō no sekai* (January 1941), 98–112.

——. "Taidan: Kankai shintaisei ron." *Jitsugyō no sekai* 38, no. 9 (September 1941), 70–77.

——. "Jū mondai o sagete, Kishi shōsō to Noyori Hideichi shachō to no taidan." *Jitsugyō no sekai* (April 1942), 68–82.

——. "Senjika no nōritsu zōshin." *Nihon nōritsu* 1, no. 1 (June 1942), 15–21.

——. "Senjika no keizai seisaku taidan (Part 1)." *Jitsugyō no sekai* (August 1942), 52–59.

——. "Kishi Nobusuke to Noyori Hideichi to no senjika no keizai seisaku taidan (Part 2)." *Jitsugyō no sekai* (September 1942), 66–75.

Kishi Nobusuke, Shiina Etsusaburō, Takeuchi Kenji et al. *Shin keizai taisei no kakuritsu o kataru zadankai sokkiroku.* Tokyo: Sekai keizai shinpōsha, 1940.

Kishi Nobusuke, Yatsugi Kazuo, and Itō Takashi, eds. "Kankai seikai rokujūnen: Daiikkai Manshū jidai." *Chūō kōron* (September 1979), 278–296.

——. "Shōkō daijin kara haisen e." *Chūō kōron* (October 1979), 286–304.

——. *Kishi Nobusuke no kaisō.* Tokyo: Bungei shunjū, 1981.

Kitagawa Kazuo. "Zaikai to kakushin kanryō." *Kaizō* (January 1941), 156–163.

Kitajō Saburō. "Shin kanryō no fujin to sono shōrai sei." *Kaizō* 17, no. 7 (July 1935), 221–228.

Kitano Shigeo. *Gunjushō oyobi gunju kaisha hō.* Tokyo: Takayama shoin, 1944.

Kiyoaki Tsuji. "Decision-Making in the Japanese Government: A Study of Ringisei." In *Political Development in Japan,* edited by Robert Ward, 457–475. Princeton: Princeton University Press, 1968.

Kiyosawa Kiyoshi. *A Diary of Darkness: The Wartime Diary of Kiyosawa Kiyoshi,* trans. Eugene Soviak and Kamiyama Tamie. Princeton: Princeton University Press, 1999.

Kobayashi Hideo. "1930 nendai Manshū kōgyōkai seisasku no tenkai katei." *Tochi seidō shigaku* 44, no. 4 (summer 1969), 19–43.

——. *"Dai-tōa kyōeiken" no heisei to hōkai.* Tokyo: Ochanomizu shobō, 1975.

——. *"Nihon kabushikigaisha" o tsukutta otoko: Miyazaki Masayoshi no shōgai.* Tokyo: Shōgakkan, 1995.

——. *Chō kanryō.* Tokyo: Tokuma shoten, 1995.

——. *Mantetsu: "Chi no shūdan" no tanjō to shi.* Tokyo: Furukawa kōbunkan, 1996.

Kokumuin sōmuchō kikakusho. *Manshūkoku keizai kensetsu ni kansuru shiryō* (June 1936).

Komagome Takeshi. *Shokuminchi teikoku Nihon no bunka tōgō.* Tokyo: Iwanami shoten, 1986.

——. "'Manshūkoku ni okeru jukyō no isō: Daidō, ōdō, kōdō." *Shisō* 851 (July 1994), 57–82.

Kornai, Janos. *The Socialist System: The Political Economy of Communism.* Princeton: Princeton University Press, 1992.

Kotkin, Steven. "Modern Times: The Soviet Union and the Interwar Conjuncture." *Kritika: Explorations in Russian and Eurasian History* 2, no. 1 (2002), 111–164.

Köttigen, Carl. "Rationalization of Industry." Speech given at Japan Industrial Club, October 28, 1929.

Koyama Sadatomo. "Seimeisen kakuritsu undō," *Manshū hyōron* (October 1931), 20–23.

——. *Manshūkoku to kyōwakai.* Dairen, Manchuria: Manshū hyōronsha, 1935.

——. *Manshū kyōwakai no hattatsu.* Tokyo: Chūō kōronsha, 1941.

Kudō Akira. "The Tripartite Pact and Synthetic Oil: The Ideal and Reality of Economic and Technical Cooperation between Japan and Germany." *Annals of the Institute of Social Science,* no. 33 (1991).

——. "The Transfer of Leading-edge Technology to Japan." *Japanese Yearbook on Business History,* no. 11 (1994).

Kurzman, Dan. *Kishi and Japan.* New York: Obolensky, 1960.

Kusayanagi Daizō. *Jitsuroku Mantetsu chōsabu.* 2 vols. Tokyo: Asahi shinbunsha, 1979.

Lange, Oskar. "On the Economic Theory of Socialism." *Review of Economic Studies,* part I, vol. 1, no. 1 (February 1936). Reprinted in *Economic Theory and Market Socialism: Selected Essays of Oskar Lange,* edited by Tadeusz Kowalik. Vermont: Edgar Elgar, 1994.

Laquer, Walter. *Fascism: A Reader's Guide.* Berkeley: University of California Press, 1976.

Lebovics, Herman. *Social Conservatism and the Middle Classes in Germany, 1914–1933.* Princeton: Princeton University Press, 1969.

Lu, David. *Agony of Choice: Matsuoka Yosuke and the Rise and Fall of the Japanese Empire, 1880–1946.* Lanham, MD: Lexington Books, 2002.

Lüdtke, Alf. "The Honor of Labor: Industrial Workers and the Power of Symbols under National Socialism." In *Nazism and German Society,* edited by David J. Crew, 66–109. London: Routledge, 1994.

Maier, Charles S. *In Search of Stability: Explorations in Historical Political Economy.* New York: Cambridge University Press, 1987.

Mannheim, Karl. *Man and Society in an Age of Reconstruction.* New York: Harcourt, Brace, 1954.

Manshū kaikōshū kankōkai, ed. *Aa Manshū: Kunitsukuri sangyō kaihatsusha no shuki.* Tokyo: Nōrin shuppan, 1965.

Manshū nichinichi shinbunsha, ed. *Nichi-Man kankei no gendai oyobi shōrai.* Manchuria: Manshū nichinichi shinbunsha, 1936.

Manshūkoku jitsugyōbu tōseika. "Nichi-Man keizai tōsei hōsaku yōkō ni kansuru ken." Minobe Yōji Documents H 17; K 2.9.

Manshūkoku kokumuin sōmuchō. Manshūkoku kanshiroku. Shinkyō, Manchuria: Kokumuin sōmuchō, 1938–1939.

Manshūkoku kokumuin sōmuchō jōhōsho. *Manshūkoku taikei,* no. 15: *Sangyō hen.* Shinkyō, Manchuria: Manshūkoku kokumuin sōmuchō jōhōsho, 1934.

——. *Manshūkoku taikei,* no. 16: *Zaisei kinyū hen.* Shinkyō, Manchuria: Manshūkoku kokumuin sōmuchō jōhōsho, 1934.

Manshūkoku kokumuin sōmuchō kikakusho, ed. *Manshūkoku keizai kensetsu ni kansuru shiryō.* Shinkyō, Manchuria: Manshūkoku kokumuin sōmuchō kikakusho, 1937.

Manshūkoku kyōwakai, ed. *Manshūteikoku kyōwakai soshiki enkakushi.* Tokyo: Fuji shuppan, 1940.

Manshūkoku seifu, ed. *Meiji hyakunenshi sōsho.* Vol. 91. *Manshū kenkoku jūnenshi.* Tokyo: Hara shobō, 1969.

Manshūkoku tsūshinsha, ed. *Manshūkoku gensei.* Shinkyō, Manchuria: Manshūkoku tsūshinsha, 1935–1939.

Manshūkokushi hensan kankōkai, ed. *Manshūkokushi.* Vol. 1. Sōron. Tokyo: Manmō dōhō engokai, 1971.

——. *Manshūkokushi.* Vol. 2. *Kakuron.* Tokyo: Manmō dōhō engokai, 1971.

Marcuse, Herbert. *One-Dimensional Man.* Boston: Beacon Press, 1964.

Marshall, Byron K. *Capitalism and Nationalism in Prewar Japan: The Ideology of the Business Elite, 1868–1941.* Stanford: Stanford University Press, 1967.

Maruyama Masao. *Thought and Behaviour in Modern Japanese Politics.* London: Oxford University Press, 1966.

Masaki Hisashi. "The Financial Characteristics of the Zaibatsu in Japan: The Old Zaibatsu and Their Closed Finance." In *The International Conference on Business History.* Vol. 3. *Marketing and Finance in the Course of Industrialization,* edited by Nakagawa Keiichirō, 33–54. Tokyo: University of Tokyo Press, 1978.

Matsui Haruo. "Shin tōa keizai ken no saininshiki." *Jitsugyō no sekai* (August 1940), 40–42.

Matsumae Shigeyoshi. "Kōdō kokubō kokkaron." *Kōgyō kumiai* (October 1941), 13–25.

Matsumoto Toshirō. "Dainiji taisenki no senji taisei kōsō ritsuan no ugoki—'Minobe Yōji bunsho' ni miru Nichi-Man-Shi keizai kyōgikai, dai Tōa kensetsu shingikai no katsudō." *Okayama daigaku keizai gakkai zasshi* 25, nos. 1–2 (May 1993), 99–123.

Matsusaka Y. Tak. "Managing Occupied Manchuria." In *Japan's Wartime Empire,* edited by Peter Duus, Ramon H. Myers, and Mark R. Peattie, 97–135. Princeton: Princeton University Press, 1996.

——. *The Making of Japanese Manchuria: 1904–1932.* Cambridge: Harvard University Asia Center, 2001.

Metzler, Mark. *Lever of Empire: The International Gold Standard and the Crisis of Liberalism in Prewar Japan.* Berkeley: University of California Press, 2006.

Meynaud, Jean. *Technocracy.* London: Faber and Faber, 1964.

Mikami Atsufumi. "Old and New *Zaibatsu* in the History of Japan's Chemical Industry: With Special Reference to the Sumitomo Chemical Co. and the Shōwa Denko Co." In *International Conference on Business History.* Vol. 6. *Development and Diffusion of Technology: Electrical and Chemical Industries,* edited by Ōkochi Akio and Uchida Hoshimi, 201–218. Tokyo: University of Tokyo Press, 1980.

Mikuriya Takashi. "Kokusaku sōgō kikan setsuoku mondai no shiteki tenkai: Kikakuin sōsetsu ni itaru seiji ryoku gaku." In Kindai Nihon kenkyūkai hen, *Nenpō kindai Nihon kenkyū.* Vol. 1. *Shōwaki no gunbu.* Tokyo: Yamakawa shuppansha, 1979.

Miller, Frank O. *Minobe Tatsukichi, Interpreter of Constitutionalism in Japan.* Berkeley: University of California Press, 1965.

Mimura, Janis. "Technocratic Visions of Empire: Technology Bureaucrats and the 'New Order for Science-Technology.'" In *The Japanese Empire in East Asia and its Postwar Legacy,* edited by Harald Feuss, 97–116. Munich: Iudicium Verlag, 1998.

Minami Manshū tetsudō kabushikigaisha keizai chōsakai. *Manshū keizai tōseisaku an. Vol. 2. Seisakuhen.* June, 1932.

——. "Daiikkai kantōgun bakuryō, keichō dankai kiroku." June 1935 (gokuhitsu).

——. *Manshū keizai tōsei hōsaku* 1, no. 1 (Dairen, Manchuria: Minami Manshū tetsudō keizai chōsakai, 1935).

Minobe Yōji. "Keizai shintaisei kanken." *Nihon hyōron* (March 1941), 12–20.

——. "Shin keizai taisei no kōsō." *Jitsugyō no sekai* 38, no. 11 (November 1941), 26–29.

——. *Senji keizai taisei kōwa.* Tokyo: Tachibana shoten, 1942.

Minobe Yōji and Aoki Kinichi. "Kankai e no yōbō: Zaikai e no yōbō." *Jitsugyō no Nihon* (June 1941), 38–45.

Mitani Taiichirō. "Manshūkoku kokka taisei to Nihon no kokunai seiji." *Kindai Nihon to shokuminchi.* Vol. 2. *Teikoku tōchi no kōzō, 179–213.* Tokyo: Iwanami kōza, 1992.

Miwa Kimitada. "Japanese Policies and Concepts for a Regional Order in Asia, 1938–1940." In *The Ambivalence of Nationalism: Modern Japan between East and West,* edited by James W. White, Michio Umegaki, and Thomas R.H. Havens, 133–156. Lanham, MD: University Press of America, 1990.

Miyamoto Takenosuke. "Tairiku hatten to gijutsu." *Kagakushugi kōgyō* (April 1938), 137–146.

———. "Gijutsu to keizai to no kadai," *Gijutsu hyōron* (September 1939), 2–10.

———. *Kagaku no dōin.* Tokyo: Kaizōsha, 1941.

———. *Tairiku no keizai kensetsu.* Tokyo: Iwanami shoten, 1941.

Miyauchi Isamu, ed. *Manshūkenkoku sokumenshi: Kenkoku jisshūnen kinen.* Tokyo: Shinkeizaisha, 1942.

Molony, Barbara. "Innovation and Business Strategy in the Prewar Chemical Industry." In *The International Conference on Business History.* Vol. 15. *Japanese Management in Historical Perspective,* edited by Yui Tsunehiko and Nakagawa Heiichirō, 141–166. Tokyo: University of Tokyo Press, 1989.

———. *Technology and Investment: The Prewar Japanese Chemical Industry.* Cambridge, MA: Harvard University Asia Center, 1990.

Moore, Barrington, Jr. *Social Origins of Dictatorship and Democracy.* Boston: Beacon Press, 1966.

Morgan, Alfred D. "The Japanese War Economy: A Review." *The Far Eastern Quarterly* 8, no. 1 (November 1948), 64–71.

Mōri Hideoto. "Shina ni taisuru keizai gijutsu no mondai" (Speech given at Tōa kenkyūjo on March 15, 1939). Tōa keizai kenkyūjo (July 1939).

———. "Taishi keizai gijutsu no kōzō." *Keizai jōhō* (June 1939), 97–105.

———. "Shina no sangyō kaihatsu" (based on speech of May 29, 1939). *Tōyō* (August 1939), 73–83.

———. "Kankai ni motomerareru mono." *Nihon hyōron* (March 1941), 54–57.

———. "Kakushin hikakushin o chōkoku suru mono." *Kaizō* (April 1941), 141–145.

———. "Dai tōa sensō to Ei-Bei sensō keizairyoku." *Kokusaku hōsō* (March 1942), 8–14.

———. "Shakai jichishugi teishō." *Mōri Hideoto kankei bunsho,* no. 230.

———. "Tairiku no kanki: Shikaisaru ni shinobinai—watakushidomo no shinjitsu na kimochi." *Mōri Hideoto kankei bunsho, Document 234.*

Mōri Hideoto and Miki Kiyoshi. "Taidankai: Ashita no kagaku Nihon no sōzō." *Kagakushugi kōgyō* (January 1941), 186–207.

Mōri Hideoto, Nagata Kiyoshi, Nakayama Ichirō, and Hatano Kanae. "Sensō zaisei o tsuku." *Kaizō* (October 1941), 212–230.

Morikawa Kakuzō. "Doitsu no gijutsusha to Nihon no gijutsusha." Minobe Yōji Documents L 1:77.

Morris-Suzuki, Tessa. *The Technological Transformation of Japan: From the Seventeenth to the Twenty-First Century.* Cambridge: Cambridge University Press, 1994.

Mutō Tomio. *Watakushi to Manshūkoku.* Tokyo: Bungei shunjū, 1988.

Myers, Ramon H. *The Japanese Economic Development of Manchuria, 1932–1945.* New York: Garland, 1982.

———. "Creating an Enclave Economy: The Economic Inte+gration of Japan, Manchuria, and North China, 1932–1945." In *Japan's Wartime Empire,* edited by Peter Duus, Ramon H. Myers, and Mark R. Peattie. Princeton: Princeton University Press, 1996.

Myers, Ramon H., and Mark Peattie, eds. *Japanese Colonial Empire, 1895–1945.* Princeton: Princeton University Press, 1987.

Najita, Tetsuo. *Japan: The Intellectual Foundations of Modern Japanese Politics.* Chicago: University of Chicago Press, 1974.

Najita, Tetsuo, and Harry Harootunian. "Japanese Revolt against the West: Political and Cultural Criticism in the Twentieth Century." In *The Cambridge History of Japan.* Vol. 6. *Twentieth Century,* edited by Peter Duus, 711–774. Cambridge: Cambridge University Press, 1988.

Nakagawa Keiichirō. "Business Strategy and Industrial Structure in Pre–World War II Japan." In *The International Conference on Business History*. Vol. 1. *Strategy and Structure of Big Business*, edited by Nakagawa Keiichirō, 3–38. Tokyo: University of Tokyo Press, 1976.

Nakamura Takafusa. *Nihon no keizai tōsei: Senji, sengo no keiken to kyōkun*. Tokyo: Nihon keizai shinbun, 1974.

Nakamura Takafusa and Hara Akira. "Keizai shintaisei." In *"Konoe shintaisei" no kenkyū*, edited by Nihon seiji gakkai, 71–133. Tokyo: Iwanami shoten, 1972.

Nakamura, Takafusa, Itō Takashi, and Hara Akira eds. *Gendaishi o tsukuru hitobito*. Vols. 1–3. Tokyo: Mainichi shinbunsha, 1971–1972.

Nakamura Takafusa and Miyazaki Masayasu, eds. *Kishi Nobusuke seiken to kōdō seichō*. Tokyo: Tōyō keizai shinpōsha, 2003.

Nichi-Man jitsugyō kyōkai. *Kensetsu tojō no Manshūkoku*. Tokyo: Nichi-Man jitsugyō kyōkai, May 1934).

——. *Manshūkoku keizai seisaku no genzai to shōrai ni tsuite*. Tokyo: Manshū jitsugyō kyōkai.

Nichi-Man zaisei keizai kenkyūkai hen. *Nachisu keizai hō*. Tokyo: Nihon hyōronsha, 1937.

Nihon hyōron shinsha, ed. *Yōyōtaru: Minobe Yōji tsuitōroku*. Tokyo: Nihon hyōron shinsha, 1954.

Nihon kagakushi gakkai, ed. *Nihon kagaku-gijutsushi taikei*. Vol. 4. Tokyo: Daiichi hōki shuppan, 1966.

Nihon kokusai seiji gakkai taiheiyō sensō gen'in kenkyūbu, ed. *Taiheiyō sensō e no michi*. Vol. 8. *Shiryō bekkan*. Tokyo: Asahi shinbunsha, 1962–1963.

Noguchi Yukio. *1940-nen taisei*. Tokyo: Tōyō keizai shinbunsha, 1995.

Nolan, Mary. *Visions of Modernity*. Oxford: Oxford University Press, 1994.

Norman, E. Herbert. "The Genyosha: A Study in the Origins of Japanese Imperialism." *Pacific Affairs* 17, no. 3 (September 1944), 261–284.

Notar, Ernest J. "Japan's Wartime Labor Policy: A Search for a Method," *Journal of Asian Studies* 44, no. 2 (February 1985), 311–328.

Ōbayashi Shinji. "Keizaitetsugaku hōhō ron." In *Kindai Nihon keizai shisōshi*. Vol. 2. Edited by Chō Yukio and Sumiya Kazuhiko, 219–250. Tokyo: Yūhikaku, 1971.

Ogata Sadako N. *Defiance in Manchuria: The Making of Japanese Foreign Policy, 1931–1932*. Berkeley: University of California Press, 1964.

Ogimura Ryūtarō. "Denryokuan hantai no nami." *Keizai ōrai* (October 1936), 314–320.

Okazaki Tetsuji. "The Japanese Firm under the Wartime Planned Economy." In *The Japanese Firm: The Sources of Competitive Strength*, edited by Masahiko Aoki and Ronald Dore, 350–378. Oxford: Oxford University Press, 1994.

——. "The Wartime Institutional Reforms and Transformation of the Economic System." In *The Political Economy of Japanese Society*. Vol. 1. *The State or the Market?* edited by Banno Junji, 277–345. Oxford: Oxford University Press, 1997.

Ōkōchi kinenkaihen. *Ōkōchi Masatoshi: Hito to sono jigyō*. Tokyo: Nikkan kōgyō shinbunsha, 1954.

Ōkōchi Masatoshi. "Riken kontsuerun no shimei." *Kagakushugi kōgyō* (June–July, 1937), 60–72.

——. "Seisan ryoku kakujū to jikkyū jissoku." *Kagakushugi kōgyō* (July 1938), 2–7.

——. "Shihon to keiei no bunri." *Kagakushugi kōgyō* (January 1941), 10–22.

Okumura Katsuko, ed. *Tsuioku Okumura Kiwao*. Tokyo: Okumura Katsuko, 1970.

Okumura Kiwao. *Yūbin hō ron*. Tokyo: Katsumeidō shoten, 1927.

——. *Teishin ronsō*. Tokyo: Kōtsū kenkyūsha, 1935.

——. *Denryoku kokuei*. Tokyo: Kokusaku kenkyūkai, 1936.

——. *Nihon seiji no kakushin.* Tokyo: Ikuseisha, 1938.

——. *Henkakuki Nihon seiji keizai.* Tokyo: Sasaki shobō, 1940.

——. "Kakushin to wa nanizoya." *Jitsugyō no Nihon* (January 1941), 44–51.

——. "Haisen Beikoku no kokumin seikatsu." *Shūhō* (June 1942), 22–30.

——. "Henkanki to seinen no shimei." *Jitsugyō no Nihon* (July 1941), 38–43.

——. "Shisōsen to kagaku." *Shūhō* (October 1942), 20–29.

——. "Kagayakashiki sekai shin chitsujo." *Kokusaku hōsō* (March 1943), 1–11.

——. *Sonnō jōi no kessen.* Tokyo: Ōbunsha, 1943.

——. "Denryoku kokkan mondai." Interview in Andō Yoshio, ed. *Shōwashi e no shōgen.* Vol. 3, 149–165. Tokyo: Hara shobō, 1993.

Ono Eiji. "Shinkō zaibatsu no shisō." In *Kindai Nihon keizai shisō.* Vol. 2. Edited by Chō Yukio and Sumiya Kazuhiko. Tokyo: Yūhikaku, 1971.

Orwell, George. "Second Thoughts on James Burnham," *Polemic* 3 (May 1946), 13–33.

Ōtani Keijirō. *Gunbatsu: Ni-ni-roku jiken kara haisen made.* Tokyo: Tosho shuppansha, 1971.

Ōyodo Shōichi. *Miyamoto Takenosuke to kagaku gijutsu gyōsei.* Tokyo: Tōkai daigaku shuppansha, 1989.

——. *Gijutsu kanryō no seiji sankaku.* Tokyo: Chūō shinsho, 1997.

Ozaki Yukio. "Kanryō ron." *Nihon hyōron* 11, no. 11 (November 1936).

Pauer, Erich. "Japan's Technical Mobilization in the Second World War." In *Japan's War Economy,* edited by Erich Pauer, 39–64. London: Routledge, 1999.

——, ed. *Japan's War Economy.* London: Routledge, 1999.

Paxton, Robert O. *The Anatomy of Fascism.* New York: Alfred A. Knopf, 2004.

Payne, Stanley G. "Fascism, Nazism, and Japanism," *International History Review* 6, no. 2 (May 1984), 265–276.

Peattie, Mark R. *Ishiwara Kanji and Japan's Confrontation with the West.* Princeton: Princeton University Press, 1975.

Polanyi, Karl. *The Great Transformation.* Boston: Beacon Press, 1944.

Pyle, Kenneth B. "The Advantages of Followership: German Economics and Japanese Bureaucrats, 1890–1925." *Journal of Japanese Studies* 1, no. 1 (autumn 1974), 127–164.

Reynolds, E. Bruce. *Japan in the Fascist Era.* New York: Palgrave, 2004.

Rice, Richard. "Economic Mobilization in Wartime Japan: Business, Bureaucracy, and Military in Conflict." *Journal of Asian Studies* 38, no. 4 (August 1979), 689–706.

Rinji sangyō chōsakyoku. "Kakkoku sangyō tōseihō no shiteki hattatsu" (October 1934). Minobe Yōji Documents, H 21; K 2.5.

Rinji sangyō gōrikyoku. "Rinji sangyō gōrikyoku no jigyō." Tokyo: Rinji sangyō gōrikyoku, 1935.

Rollins, William H. "Whose Landscape" Technology, Fascism, and Environmentalism on the National Socialist Autobahn." *Annals of the Association of American Geographers* 85, no. 3 (September 1995), 494–520.

Rōyama Masamichi. *Japan's Position in Manchuria.* The Japan Council, Institute of Pacific Relations, 1929.

——. "Seiji." In *Manmō jijō sōran,* 81–132. Tokyo: Kaizōsha, 1932.

——. "Gijutsu to gyōsei." *Kagakushugi kōgyō* (May 1938), 13–23.

Ryū Shintarō. *Nihon keizai no saihensei.* Tokyo: Chūō kōron, 1939.

Saaler, Sven, and J. Victor Koschmann, eds. *Pan-Asianism in Modern Japanese History: Colonialism, Regionalism, and Borders.* London: Routledge, 2007.

Sakisaka Itsurō, ed. *Nihon tōsei keizai zenshū.* Vols. 1–10. Tokyo: Kaizōsha, 1933–1934.

Sakomizu Hisatsune. *Kaisha rieki haitōrei gaisetsu.* Tokyo: Ōkurasho, 1939.

——. "Kaisha keiri tōseirei o kataru." *Jitsugyō no Nihon* (April 1940), 14–23.

——. "Keizai tōsei no shin dankai." *Jitsugyō no Nihon* (June 1941), 46–48.

——. "Senjika no kabushiki mondō." *Jitsugyō no Nihon* (November 1941), 28–33.

——. *Zaisei kinyū kihon hōsaku kaisetsu.* Tokyo: Senji seikatsu sōdansho, 1941.

——. *Kinyū tōseikai no shinro.* Tokyo: Shinkeizaisha, 1942.

——. "Kokka sōryokusen to zaisei." *Tōyō keizai shinpō* (April 1943), 12–17.

——. Interview. In *Gendaishi o tsukuru hitobito.* Vol. 3, edited by Nakamura Takafusa, Itō Takashi, and Hara Akira, 45–105. Tokyo: Mainichi shinbunsha, 1971.

Samuels, Richard J. *The Business of the Japanese State.* Ithaca: Cornell University Press, 1987.

——. *Rich Country, Strong Army: National Security and the Technological Transformation of Japan.* Ithaca: Cornell University Press, 1994.

——. *Machiavelli's Children: Leaders and Their Legacies in Italy and Japan.* Ithaca: Cornell University Press, 2003.

Sarfatti-Larson, Magali. "Notes on Technocracy: Some Problems of Theory, Ideology, and Power." *Berkeley Journal of Sociology* 17, no. 5 (1972–1973), 1–34.

Sasaki Tsutomu. "Sendoteki handō to modanizumu no ketsuki no shōso." *Shisō,* no. 805 (July 1991).

Sawai Minoru. "Kagaku gijutsu shintaisei kōsō no tenkai to gijutsuin no tanjō." *Osaka daigaku keizaigaku* 41, nos. 2–3 (December 1991), 367–395.

——. "Nitchū sensōki no kagaku gijutsu seisaku." In *Nenpō kindai Nihon kenkyū.* Vol 13. Edited by Kindai Nihon kenkyū kai, 175–197. Tokyo: Yamakawa shuppansha, 1991.

——. "Policies for the Promotion of Science and Technology in Wartime Japan." *Keizaigaku ronshū* 35, no. 1 (June 1995), 44–64.

Schumpeter, Elizabeth B. *The Industrialization of Japan and Manchukuo, 1930–1940.* New York: Macmillan, 1940.

Scott, James C. *Seeing Like a State: How Certain Schemes to Improve the Human Condition Have Failed.* New Haven: Yale University Press, 1998.

Selznick, Philip. *TVA and the Grass Roots: A Study in the Sociology of Formal Organization.* New York: Harper Torchbooks, 1966.

Shiina Etsusaburō. "Dokusen kigyō no zehi." *Jitsugyō no sekai* (December 1940), 66–67.

——. *Senji keizai to busshi chōsei.* Tokyo: Sangyō keizai gakkai, 1941.

——. "Nanpō shinshutsu no kamae." *Jitsugyō no Nihon* (March 1942), 20–25.

——. "Senryoku nōritsu to nōritsu zōshin." *Nihon nōritsu* 2, no. 4 (April 1943), 6–10.

——. *Watakushi no rirekisho.* Vol. 41. Tokyo: Nihon keizai shinbunsha, 1970, 157–230.

——. Interview. In *Gendaishi o tsukuru hitobito.* Vol. 4. Edited by Nakamura Takafusa, Itō Takashi, and Hara Akira, 251–311. Tokyo: Mainichi shinbunsha, 1971.

——. "Nihon sangyō no daijikkenjō—Manshū." *Bungei shunjū* (February 1976), 106–114.

Shiina Etsusaburō, Kōno Mitsu, et al. "Shintaisei to kōgyō no saihensei o kataru zadankai." *Kōgyō kumiai* (October 1940), 29–55.

Shiina Etsusaburō tsuitōroku kankōkai, ed. *Kiroku: Shiina Etsusaburō.* 2 vols. Tokyo: Shiina Etsusaburō tsuitōroku kankōkai, 1982.

Shōwa kenkyūkai jimukyoku. "Nihon keizai saihensei shian" (August 1940). Minobe Yōji Documents G 2:16.

Silberman, Bernard S. "Ringisei—Traditional Values or Organizational Imperatives in the Japanese Upper Civil Service: 1868–1945." *Journal of Asian Studies* 32, no. 2 (February 1973), 251–264.

——. "The Bureaucratic State in Japan: The Problem of Authority and Legitimacy." In *Conflict in Modern Japanese History,* edited by Tetsuo Najita and J. Victor Koschmann, 226–257. Princeton: Princeton University Press, 1982.

Smethurst, Richard J. *A Social Basis for Prewar Japanese Militarism: The Army and the Rural Community.* Berkeley: University of California Press, 1974.

———. *From Foot Soldier to Finance Minister: Takahashi Korekiyo, Japan's Keynes.* Cambridge, MA: Harvard University Asia Center, 2007.

Smith, Henry. *Japan's First Student Radicals.* Cambridge: Harvard University Press, 1972.

Sōmuchō kikakusho. *Sōgō ritchi keikaku sakutei yōkō* (June 1940).

Spaulding, Robert, M. Jr. 1957. "Japan's 'New Bureaucrats,' 1932–45." In *Crisis Politics in Prewar Japan,* edited by George M. Wilson, 51–70. Tokyo: Sophia University, 1970.

———. "The Bureaucracy as a Political Force, 1920–1945." In *Dilemmas of Growth in Prewar Japan,* edited by James W. Morley, 33–80. Princeton: Princeton University Press, 1971.

Steinberg, John W., Bruce W. Menning, et al., eds. *The Russo-Japanese War in Global Perspective: World War Zero.* Leiden: Brill, 2005.

Sternhell, Zeev. *Neither Right nor Left: Fascist Ideology in France,* trans. David Maisel. Princeton: Princeton University Press, 1986.

———. *The Birth of Fascist Ideology.* Princeton: Princeton University Press, 1994.

Sugai Shirō. *Kokudo keikaku no keika to kadai.* Tokyo: Taimeidō, 1975.

Sugihara Masami. "Ni-Shi jiken ni kataserareta sekaishiteki kadai." *Kaibō jidai* (July 1938), 6–19.

Suzuki Teiichi. *Suzuki Teiichi shi danwa sokkiroku.* Tokyo: Nihon Nihon kindai shiryō kenkyūkai, 1971.

Sweeney, Dennis. "Corporatist Discourse and Heavy Industry in Wilhelmine Germany: Factory Culture and Employer Politics in the Saar." *Comparative Studies in Society and History* 43, no. 4 (October 2001), 701–734.

Szpilman, Chris W.A. "The Dream of One Asia: Ōkawa Shūmei and Japanese Pan-Asianism." In *The Japanese Empire in East Asia and its Postwar Legacy,* edited by Harald Fuess. Munich: Iudicium Verlag, 1998.

———. "Kita Ikki and the Politics of Coercion." *Modern Asian Studies* 36, no. 2 (2002), 467–490.

Tachibana Shiraki. "Manshū jihen to fuashizumu." *Manshū hyōron* (November 1931), 2–9.

Taikakai (Gotō Fumio), ed. *Naimushōshi.* Vol. 1. Tokyo: Chihō zaimu kyōkai, 1971.

Taiyōji Jun'ichi. "Die geistigen Grundlagen der industriellen Entwicklung in Japan." In *Die industrielle Entwicklung in Japan unter besonderer Berücksichtigung seiner Wirtschafts—und Finanzpolitik: Schriftenreihe zur Industrie- und Entwicklungspolitik.* Vol. 1. Edited by Ikeda Kōtarō, Katō Yoshitarō, and Taiyōji Jun'ichi, 167–228. Berlin: Duncker und Humblot, 1970.

Tajiri Ikuzō. *Shōwa no yōkai: Kishi Nobusuke.* Tokyo: Gakuyō shobō, 1979.

Takahashi Hikohiro. "Shin kanryō, kakushin kanryō, shakaiha kanryō: Kyōchōkai bunseki no isshikaku toshite." *Shakai rōdō kenkyū* 43, nos. 1–2 (November 1996), 33–64.

Takahashi Masae, ed. *Gendaishi shiryō.* Vol. 5. *Kokkashugi undō.* Tokyo: Misuzu shobō, 1964.

Tanaka Shinichi. "Sangyō ni okeru 'shintaisei' an yōkō (sōan)" Minobe Yōji Documents G 2:14.

———. *Nihon sensō keizai hishi.* Tokyo: Computer Age, 1974.

Tansman, Alan, ed. *The Culture of Japanese Fascism.* Durham, NC: Duke University Press, 2009.

Tilman, Rick. *Thorstein Veblen and His Critics, 1891–1963.* Princeton: Princeton University Press, 1992.

Touraine, Alain. *The Post-Industrial Society.* London: Wildwood House, 1971.

Tsūshō sangyōsho. *Shōkō seisakushi.* Vol. 11. *Sangyō tōsei.*Tokyo: Shōkō seisakushi kankōkai, 1964.

Tsutsui Kiyotada. *Shōwaki Nihon no kōzō.* Tokyo: Yūhikaku, 1985.

Tsutsui, William M. *Manufacturing Ideology: Scientific Management in Twentieth-Century Japan.* Princeton: Princeton University, 1998.

Udagawa Masaru. *Shinkō zaibatsu.* Tokyo: Nihon keizai shinbunsha, 1984.

Udagawa Masaru and Seishi Nakamura. "Japanese Business and Government in the Interwar Period: Heavy Industrialization and the Industrial Rationalization Movement." In *The International Conference on Business History.* Vol. 5. *Government and Business,* edited by Nakagawa Keiichirō, 83–100. Tokyo: University of Tokyo Press, 1980.

Uno Kazushiro. "Ayukawa, Mori, Nakano, Noguchi." *Jitsugyō no Nihon* (January 1938), 166–169.

Ushio Shiota. *Kishi Nobusuke.* Tokyo: Kōdansha, 1996.

Veblen, Thorstein. *The Engineers and the Price System.* New York: Viking Press, 1921.

——. *What Veblen Taught.* Edited by Wesley C. Mitchell. New York: Viking Press, 1936.

——. "The Opportunity of Japan." *Essays in Our Changing Order.* New York: Viking Press, 1952.

Wada Hidekichi. "Kaiketsu Ayukawa Yoshisuke." *Nihon hyōron* (July 1936), 421–438.

Wada Hidekichi, Baba Tsunego, Rōyama Masamichi et al. "Kanryō no chōryō to fuhai o kataru zadankai." *Nippon hyōron* 11, no. 11 (November 1936), 24.

Wakamiya Yoshibumi. *The Postwar Conservative View of Asia.* Tokyo: LTCB International Foundation, 1997.

Weiner, Susan Beth. "Bureaucracy and Politics in the 1930s: The Career of Gotō Fumio." Ph.D. Diss., Harvard University, 1984.

Weisskopf, Walter A. "Same Old New Class." *New York Review of Books,* 9, no. 10 (December 1967).

Wilson, George, M. *Radical Nationalist in Japan: Kita Ikki, 1883–1937.* Cambridge, MA: Harvard University Press, 1969.

Wiskemann, Erwin and Heinz Lütke, *Der Weg der deutschen Volkswirtschaftslehre: Ihre Schöpfer und Gestalten* (Berlin: Junker & Dünnhaupt Verlag, 1937).

Wolf, Kurt H. *From Karl Mannheim.* New York: Oxford University Press, 1971.

Wolferen, Karel van. *The Enigma of Japanese Power.* London: Macmillan, 1989.

Woolf, S. J., ed. *The Nature of Fascism.* London: Weidenfeld & Nicolson, 1968.

Wray, William D. "Asō Hisashi and the Search for Renovation in the 1930s." Papers on Japan, Vol. 5. Cambridge, MA: Harvard East Asia Research Center, 1970, 55–98.

Yabe Teiji. "Konoe Fumimaro to shintaisei." In *Kindai Nihon o tsukatta hitobito.* Vol. 1. Edited by Ōkōchi Kazuo and Ōya Sōichi. Tokyo: Mainichi shinbunsha, 1965.

Yagi Kiichirō. "Economic Reform Plans in the Japanese Wartime Economy: The Case of Shintarō Ryū and Kei Shibata." In *Economic Development in Twentieth Century East Asia: The International Context,* edited by Aiko Ikeo. London: Routledge, 1997.

Yamaguchi Jūji. *Kieta teikoku Manshū.* Tokyo: Mainichi shinbunsha, 1967.

Yamamuro Shinichi. *Manchuria under Japanese Dominion,* trans. Joshua A. Fogel. Philadelphia: University of Pennsylvania Press, 2006.

——. "'Manshūkoku' tōchi kateiron." In *"Manshūkoku" no kenkyū,* edited by Yamamoto Yūzō, 83–129. Tokyo: Ryokuin shobō, 1995.

Yamanouchi Yasushi, J. Victor Koschmann, and Ryūichi Narita, eds. *Total War and "Modernization."* Ithaca: Cornell East Asia Series, 1998.

Yanagida Kōmine, Matsui Haruo, Katō Kōgorō, Ōkōchi Masatoshi, and Ishibashi Tanzan. "Nihon no shigen o kataru kai." *Kagakushugi kōgyō* (June 1939), 164–182.

Yanagisawa Osamu. "The Impact of German Economic Thought on Japanese Economists before World War II." In *The German Historical School: The Historical and Ethical Approach to Economics,* edited by Yuichi Shionoya, 173–187. London: Routledge, 2001.

——. "'Gemeinnutz geht vor Eigennutz' im Streit um die Neue Wirtschaftsordnung in Japan in der kritischen Zeit," in Rainer Gömmel and Markus A. Denzel, eds., *Weltwirtschaft und Wirtschaftsordnung: Festschrift für Jürgen Schneider zum 65 Geburtstag,* 301–314. Stuttgart: Franz Steiner Verlag, 2002.

Yatsugi Kazuo. Interview. In *Gendaishi o tsukuru hitobito.* Vol. 4, 45–140. Edited by Takafusa Nakamura, Takashi Itō, and Akira Hara. Tokyo: Mainichi shinbunsha, 1973.

——. *Shōwa dōran shishi.* 3 vols. Tokyo: Keizai ōraisha, 1978.

Yokota Shūhei. *Kokudo keikaku to gijutsu.* Tokyo: Shōkō gyōseisha, 1944.

Yomiuri shinbunsha. *Shōwa shi no tennō.* Vols. 16–18. Tokyo: Yomiuri shinbunsha, 1971.

Yoshida Kazuo. *Gottoru: Seikatsu toshite no keizai.* Tokyo: Dōbunkan, 2004.

Yoshida Yutaka. "Gunji shihai: Manshū jihen ki." In *Nihon teikokushugi no Manshū shihai: Jūgonen sensōki o chūshin ni,* edited by Asada Kyōji and Kobayashi Hideo, 93–162. Tokyo: Jichōsha, 1986.

Yoshimoto shigeyoshi. *Kishi Nobusuke den.* Tokyo: Tōyō shokan, 1957.

Yoshino shinji. *Wagakuni kōgyō no gōrika.* Tokyo: Nihon hyōronsha, 1930.

——. *Nihon kōgyō seisaku.* Tokyo: Nihon hyōronsha, 1935.

——. "Keizai kokunan to sangyō tōsei." *Kōgyō kumiai* (January 1941), 4–6.

——. *Omokaji torikaji: Ura kara mita Nihon sangyō no ayumi.* Tokyo: Tsūshō sangyō kenkyūsha, 1962.

——. "Sangyō gōrika." In *Shōwa shi e no shōgen* 1, edited by Andō Yoshio, 164–184. Tokyo: Hara shobō, 1993.

Young, Louise. *Japan's Total Empire: Manchuria and the Culture of Wartime Imperialism.* Berkeley: University of California Press, 1998.

Studies of the Weatherhead East Asian Institute
Columbia University

SELECTED TITLES

Complete list at http://www.columbia.edu/cu/weai/weatherhead-studies.html.

Imperial Japan at Its Zenith: The Wartime Celebration of the Empire's 2,600th Anniversary, by Kenneth J. Ruoff. Cornell University Press, 2010.

Postwar History Education in Japan and the Germanys: Guilty Lessons, by Julian Dierkes. Routledge, 2010.

The Aesthetics of Japanese Fascism, by Alan Tansman. University of California Press, 2009.

The Growth Idea: Purpose and Prosperity in Postwar Japan, by Scott O'Bryan. University of Hawai'i Press, 2009.

National History and the World of Nations: Capital, State, and the Rhetoric of History in Japan, France, and the United States, by Christopher Hill. Duke University Press, 2008.

Leprosy in China: A History, by Angela Ki Che Leung. Columbia University Press, 2008.

Kingdom of Beauty: Mingei and the Politics of Folk Art in Imperial Japan, by Kim Brandt. Duke University Press, 2007.

Mediasphere Shanghai: The Aesthetics of Cultural Production, by Alexander Des Forges. University of Hawai'i Press, 2007.

Modern Passings: Death Rites, Politics, and Social Change in Imperial Japan, by Andrew Bernstein. University of Hawai'i Press, 2006.

The Making of the "Rape of Nanjing": The History and Memory of the Nanjing Massacre in Japan, China, and the United States, by Takashi Yoshida. Oxford University Press, 2006.

Bad Youth: Juvenile Delinquency and the Politics of Everyday Life in Modern Japan, 1895–1945, by David Ambaras. University of California Press, 2005.

Rearranging the Landscape of the Gods: The Politics of a Pilgrimage Site in Japan, 1573–1912, by Sarah Thal. University of Chicago Press, 2005.

The Merchants of Zigong: Industrial Entrepreneurship in Early Modern China, by Madeleine Zelin. Columbia University Press, 2005.

Science and the Building of a Modern Japan, by Morris Low. Palgrave Macmillan, 2005.

Kinship, Contract, Community, and State: Anthropological Perspectives on China, by Myron L. Cohen. Stanford University Press, 2005.

Reluctant Pioneers: China's Expansion Northward, 1644–1937, by James Reardon-Anderson. Stanford University Press, 2005.

Takeuchi Yoshimi: Displacing the West, by Richard Calichman. Cornell East Asia Program, 2004.

Gutenberg in Shanghai: Chinese Print Capitalism, 1876–1937, by Christopher A. Reed. UBC Press, 2004.

Japan's Colonization of Korea: Discourse and Power, by Alexis Dudden. University of Hawai'i Press, 2004.

Divorce in Japan: Family, Gender, and the State, 1600–2000, by Harald Fuess. Stanford University Press, 2004.

Japan's Imperial Diplomacy: Consuls, Treaty Ports, and War with China, 1895–1938, by Barbara Brooks. University of Hawai'i Press, 2000.

Assembled in Japan: Electrical Goods and the Making of the Japanese Consumer, by Simon Partner. University of California Press 1999.

Civilization and Monsters: Spirits of Modernity in Meiji Japan, by Gerald Figal. Duke University Press, 1999.

Bicycle Citizens: The Political World of the Japanese Housewife, by Robin LeBlanc. University of California Press, 1999.

Alignment despite Antagonism: The United States, Japan, and Korea, by Victor Cha. Stanford University Press, 1999.

Japan's Foreign Policy after the Cold War: Coping with Change, by Gerald L. Curtis, ed. M. E. Sharpe, 1993.

Index

advanced national defense state (*kōdō kokubō kokka*), 3, 30, 148, 167, 183; Manchukuo as an, 70; Okumura's definition of, 125, 144–145; and Pacific War, 170, 173, 177, 180
agrarianism, 4, 45, 108
Ah Manchuria, 199
Aikawa Haruki, 129, 160, 163
Aikawa Shūji, 160, 161
Akamatsu Katsumaro, 146–147
Akinaga Tsukizō, 15, 16, 35, 78, 91–93, 95, 150–151, 172, 173, 181–182, 191
Akita Kiyoshi, 130
Akiyama Teisuke, 129
Amakasu Masahiko, 72, 84
Aoki Kazuo, 78
Araki Sadao, 45
Arisawa Hiromi, 35
army, and Ayukawa, 88, 102; conspiracies, 41–45, 47; and Manchukuo, 48, 50, 71, 75, 78, 96; and Munitions Ministry, 181; pamphlet, 44–45, 68; and propaganda, 185; and reform bureaucrats, 33, 69, 70, 118, 122–123, 126, 128, 147; technocrats in, 2, 9, 14, 15–17; and technology policy, 159; total war planning, 16–21, 17n26, 85, 141, 146; traditional view of, 15. *See also* Control faction; Imperial Way faction; Kwantung Army
Ashida Hitoshi, 197
Asia Development Board (Kōain), 72, 109, 129, 132, 132n99, 142, 157, 158
Asia Development Group (Kōa dantai), 132
Asō Hisashi, 45, 129, 136, 146, 199
autarky, 3, 17, 17n26, 18, 21, 27, 64, 95, 109, 112, 136, 187
Automobile Manufacturing Law (1936), 66, 88
Ayukawa Yoshisuke, 2, 21–28, 67–68, 88–89, 94, 100, 102–105, 174, 175

Baba Eichi, 140
Bank of Japan, 156
Banks and Other Financial Institutions Funds Utilization Ordinance, 155
Berger, Gordon, 147
Berle, Adolf, 121
Brady, Robert, 38, 109

bureaucracy, and big business, 22, 136, 164, 168–169; culture of, 16, 29–30, 175, 179; and the managerial state, 29–30; of Manchukuo, 55, 62, 75, 76–80, 106; and modernity, 10; new order for, 172, 177–181; postwar, 199; structure of, 14, 20, 29, 32, 40; technocrats and, 6, 7–8, 13–14, 29, 34, 40, 132, 158, 168–169; traditional conception of, 29, 195, 198. *See also* reform bureaucrats
Burnham, James, 7–8
business, and army; 19, 21, 41, 45, 46, 59–61; critique of new order, 136, 143, 153–154, 158, 163, 164–169; fascist view of, 61, 108, 113, 116, 165; ideology, 23, 26–29, 153; Kishi and, 2, 34, 35–39, 69, 85, 91, 97–102, 139, 150–154, 165, 172, 173–176, 179–180, 187, 189, 199; and laissez-faire approach of, 1, 29, 158, 165, 195; Manchukuo, 59–61, 65, 90–94, 97–105, 128; and Pacific War, 21–22, 170, 172, 191, 195; resistance to state control, 69, 88, 96, 119, 121–123, 140, 142, 145; and right wing, 41, 46; 164, 165, 172; technocratic view of, 1, 30–31, 40, 70, 139, 164, 167–168; zaibatsu groups, 21–29, 40, 45, 59–65, 67, 103–105
Business Department, 74, 79–80, 90, 94, 98

Cabinet Advisory Council, 181
Cabinet Deliberation Council, 67, 69, 85, 119
Cabinet Information Bureau, 129n82, 185–186
Cabinet Investigation Bureau, 77
Cabinet Planning Agency, 20, 119
Cabinet Planning Board, drafts for new order, 152, 157–158, 161–162, 167–168, 191; formation of, 20, 119; incident, 138, 154; members of, 2, 32, 72, 107, 105, 109, 138, 150–151; during Pacific War, 172–175, 180–181; and postwar, 198; Science Division, 157, 158–161; war mobilization, 130, 141–142, 154n51
Cabinet Planning Board Deliberation Room, 150, 167–168
Cabinet Research Bureau, 66–69, 119, 162, 177
capitalism, crisis of, 7–8, 41, 61; and fascism, 5, 52, 61, 115–116, 130, 196; liberal capitalism, 1, 8, 40; Manchuria policy on, 1, 48, 51, 61, 63, 67, 89, 91–92, 103; Marxist view of, 22, 52;

political economy, 113–116; techno-fascism, 2, 3–6, 8, 69, 107–108, 109, 112, 116, 124, 137, 172, 194, 196–197; and *tenkō*, 109, 137; as "third way," 5, 35, 40, 122, 172, 194; wartime strategy toward, 177. *See also* totalism

February incident, 41–44, 69, 122

Finance Department (Zaiseibu), 71, 80

Finance Ministry, in Manchukuo, 56n31, 71, 80, 96; and new order, 154–156, 175; during Pacific War, 176; technocrats and, 16, 32, 33, 71, 127, 128; and wartime controls, 140–142, 178

Five Year Plan, 159; Manchurian, 93, 94–97, 98, 99, 100, 101, 141, 192; Soviet, 61, 63, 96, 120, 192

Fordism, 28, 116

Foreign Exchange Control Law (1933), 140

Fourth Division, 57, 75

Fuel Bureau, 142

Fuji Electric Company, 122

Fujiwara Ginjirō, 181

functionalism, 31, 68; functionalist society, 37, 127, 136, 139, 148–149, 150, 161, 163, 169, 186, 189, 194

Funds Accommodation Ordinance, 142

Funds Bureau, 142

Funk, Walter, 181

Furumi Tadayuki, 71–72, 74, 77–78, 84, 91, 96, 128

Fushun, coal mines, 100–101, 105; oil-extraction, 87

Gaullism, 198

General Affairs Agency (Sōmuchō), 74; function of, 55–57, 57n32, 75, 77–78, 81; officials of, 71, 72, 77, 82, 94, 96, 187

geopolitics, 3, 5, 46, 126–129, 131, 137, 172, 184, 186, 189–191, 194; and Manchuria, 49–50, 127

Germany, 3, 4, 5, 7, 14, 15, 16–19, 33, 36–38, 133; and technology, 109–113, 198. *See also* Haushofer; Hitler; Nazism

Godō Takuo, 150

Gondo Seikei, 16, 54

Gotō Ryūnosuke, 68`

Gottl-Ottlilienfeld, Friedrich von, 114, 115–116, 129, 134, 153, 153n47

Government Affairs Board (Kokumuin), 55, 56

Greater East Asia Co-Prosperity Sphere, 2, 3, 64, 112, 160, 171, 199; concept of, 186, 189, 194; geopolitical vision of, 189, 191; national land planning, 191–194; resources of, 187–188

Great Japan Industrial Patriotic Association, 163

Great Japan Patriotic Press Association, 186

Great Majestic Peak Association (Daiyūhōkai), 51. *See also* Kasagi Yoshiaki

Great Unity Academy (Daidō gakuin), 74

Grossraumwirtschaft, 172, 186–188, 189

hakkō ichiu, 190

Hamada Kunimatsu, 123

Hamaguchi Osachi, 41, 67, 86

Hanaya Tadashi, 59

Hara, Akira, 97

Harada Kumakichi, 77

Harada Matsuzō, 73

Haushofer, Karl, 16, 49, 129n81

Hashimoto Kingorō, 47

Hayashi Senjūrō, 45, 75, 123

Hiranuma Kiichirō, 164

Hirota cabinet, 122–123

Historical School (German), 114

Hitler, Adolf, 4, 39n58, 113

Home Ministry, 33, 67, 117, 129n82, 147–148, 162, 164, 185, 191, 196–197

Honjō Shigeru, 71

Hoshino Naoki, 2, 3, 32, 71–73, 76, 91–94, 95–96, 102, 128, 138, 146, 150, 153, 187, 191

House of Peers, 43

hydroelectric power, 23, 24, 76, 121

Ikeda Seihin, 142

Ikeda Sumihisa, 15, 16, 35, 44–45, 68

Imada Shintarō, 59

Imperial Rule Assistance Association, 80, 129n80, 148

Imperial Way faction (Kōdō-ha), 15–16, 42, 44, 45

Important Industries Control Law (Japan, 1931), 38–39, 66, 85, 88

Important Industries Control Law (Manchuria, 1937), 97–99

Imports and Foreign Exchange Control Ordinance (1937), 140

Inaba Hidezō, 32, 33n78, 34, 154

Indochina, 171, 190

Industrial Bank of Japan, 25, 142, 155–156

Industrial Patriotic Association, 162

Industrial Patriotic movement, 162–163

industrial rationalization, 11, 28, 35–39, 67, 85–86, 133, 196; German, 36–38, 109–110, 153

Information Section, 81

link system, 143
List, George Friedrich, 38, 86–87, 134
living space, 3, 49, 50, 133, 186, 190
London Naval Treaty, 41
Ludendorff, Erich, 16, 18

Maeda Yonezō, 122
Major Export Industries Association Law, 36
managerialism, 6, 14, 61, 197; managerial revo-
    lution, 7–8
Manchukuo, 55, 67–68, 75, 81, 194, 196,
    199–200; as antithesis of Japan, 49; and
    army's economic plans, 59–65, 94–97,
    100–102; as bureaucratic training ground,
    8, 33, 71, 74, 87, 106, 173, 187; and fascism,
    8, 52, 61, 70, 106, 196; government, 55–59,
    76–78, 80–84; and Japanese imperialism, 49;
    justification for, 48, 52; leaders of, 1–2; as a
    model for Japan, 87–88, 106, 153, 177; new
    zaibatsu and, 67–68, 76–77, 103–107; opium
    trade, 128; pan-Asianist visions of, 48–49,
    51–52, 54, 55, 58–59; propaganda, 80–82;
    reform bureaucrats and, 71–75, 80–84, 87,
    89–94, 97–102, 118–119, 128, 132, 133, 171,
    173, 187, 199–200; as subordinate to Japan,
    55, 70, 80–81, 95
Manchukuo Wire Service (Manshūkoku
    tsūshinsha), 81
Manchuria, 21, 32, 44, 49, 59; arrival of Japanese
    bureaucrats in, 70–74, 72n6; attracting busi-
    ness to, 89–94, 97–105; Chinese in, 51–55, 56,
    81; economy of, 61–65, 94–97; ethnic diver-
    sity of, 54; invasion of, 42, 45, 47–48; Japanese
    in, 51–52, 59; Koreans in, 52, 55; political
    battles over planning, 75–80, 82–83; reform
    movement and, 42, 48; resource development
    in, 87–88; and rise of technocrats, 1–3, 16, 30,
    32, 33, 67–69, 105–107, 127; as strategic base,
    18, 49–50
Manchurian Affairs Bureau, 78–79, 96
Manchurian Coal Company, 100–101
Manchurian Film Society (Manshū eiga kyōkai),
    81
Manchurian Five Year Plan, Manchuria, 93,
    94–97, 98, 99–101, 141, 192
Manchurian Heavy Industries Corporation
    (Mangyō). See Mangyō
Manchurian incident, 185. See also Manchuria;
    Kwantung Army
Manchurian Light Metals, 105
Manchurian Petroleum Company, 87–88
Manchurian Public Information Society (Man-
    shū kōhōkai), 81
Manchurian Review (Manshū hyōron), 51

Manchurian Telegraph and Telephone Com-
    pany, 64, 81, 118
Manchurian Youth League, 51, 54, 58
Mangyō (Manchurian Heavy Industries Corpo-
    ration), 2, 68, 101, 102, 103–105
Mantetsu (South Manchuria Railway Corpora-
    tion), 1, 2, 42, 64, 87, 95, 96, 97, 102; advisors
    to Kwantung Army, 51–52, 57n34, 62, 62n54,
    70, 84, 100–101; conflict with bureaucrats,
    76, 77, 79; and Mangyō, 102, 105
March incident, 43, 47
Marco Polo Bridge incident, 144
Marcuse, Herbert, 198
Maruyama Masao, 4, 108
Marxism, 14; army and, 16, 35; Marxist critique
    of zaibatsu, 21–22; reform bureaucrats and,
    34, 35; technocratic critique of, 7, 10, 40
materials mobilization plans, 141, 154n51
Matsuda Reisuke, 71, 72, 95, 102
Matsui Haruo, 20, 68, 78
Matsukata zaibatsu, 23
Matsumae Shigeyoshi, 67, 184
Matsuoka Yōsuke, 2, 49, 50, 171
May 15th incident, 43
Mazaki Jinzaburō, 45
Means, Gardiner, 121
Meiji Restoration, 26, 134
middle class, 3, 14, 28, 69, 107, 108, 197, 199
Miki Kiyoshi, 129, 130, 160
Military Affairs Bureau, 45, 68, 78
Minami Iwao, 162–163
Minami Jirō, 82
Ministry of Agriculture and Forestry, 178, 181,
    191
Ministry of Commerce and Industry, 32, 33, 76,
    85, 96, 141, 142, 150, 153, 198; administrative
    reform, 177–178, 181; industrial control, 66,
    140–143, 150–154, 164, 172–175; industrial
    rationalization, 35–39; officials in Manchuria,
    69, 72, 76, 79
Ministry of Communications, 33, 117, 164, 181;
    electric power law, 66, 122; and Manchuria,
    67, 81
Ministry of Finance, 32, 33, 96, 183; financial
    control, 140, 142, 154–157, 176, 178; officials
    in Manchuria, 71, 127–128
Ministry of International Trade and Industry
    (MITI), 106, 198
Ministry of Munitions, 177, 181, 198
Ministry of Railways, 159, 181
Ministry of Welfare, 142, 175, 178
Minobe Yōji, 3, 32, 34, 35, 72, 140, 142–143, 145,
    191; administrative reform, 177–178; new
    order for industry, 150–151, 163, 164–166

www.ingramcontent.com/pod-product-compliance
Lightning Source LLC
Chambersburg PA
CBHW030648270326
41929CB00007B/261